Solutions Manual *for*

SURVEYING:
Principles and Applications
Fourth Edition

Barry F. Kavanagh

Solutions Manual for

SURVEYING: PRINCIPLES AND APPLICATIONS

Fourth Edition

BARRY F. KAVANAGH

Seneca College

Prentice Hall

Englewood Cliffs, New Jersey   Columbus, Ohio

© 1996 by Prentice-Hall, Inc.
A Simon & Schuster Company
Englewood Cliffs, New Jersey 07632

All rights reserved. Instructors of classes using Kavanagh, *Surveying: Principles and Applications, Fourth Edition,* may reproduce material from the instructor's manual for classroom use. Otherwise, no part of this book may be reproduced, in any form or by any means, without permission in writing from the publisher.

Printed in the United States of America

10 9 8 7 6 5 4 3 2

ISBN: 0-13-438318-4

CONTENTS

CHAPTER 2	PAGE 1
CHAPTER 3	PAGE 4
CHAPTER 4	PAGE 10
CHAPTER 6	PAGE 16
CHAPTER 7	PAGE 25
CHAPTER 8	PAGE 28
CHAPTER 9	PAGE 29
CHAPTER 10	PAGE 30
CHAPTER 11	PAGE 41
CHAPTER 13	PAGE 44
CHAPTER 14	PAGE 45
CHAPTER 15	PAGE 48
CHAPTER 16	PAGE 51

PREFACE

A publication of this type seems to always retain a few mistakes that have eluded detection through a number of conscientious checks. If you find any mistakes in the manual and forward the information to me - c/o Prentice Hall Inc., Englewood Cliffs, N.J., 07632, I will ensure that the corrections are sent to all those who have adopted the text.

As in the past, your comments, corrections and suggestions both for the Solutions Manual and for the text are appreciated.

Barry Kavanagh
Seneca College
1995

SOLUTIONS
CHAPTER 2

2.1 a) 20.61 x 66 = 1320.26 ft.   b) 2.18 x 66 = 143.88 ft.
    c) 23.17 x 66 = 1529.22 ft.   d) 1.60 x 66 = 105.60 ft.

2.2 a)-when looking for survey evidence to begin a survey
     -when rough-checking a construction layout
    b)-auto: identifying rural fence corners
     -wheel: frontage measurements for assessment purposes
     -wheel: measurements for accident surveys
    c)-when taking a baseline measurement
     -when measuring over long distances,or difficult terrain
    d)-when performing a topographic survey
     -when performing a shore-based hydrographic survey
    e)-when measuring across a busy highway
     -when setting a calibration baseline for industrial use
    f)-when doing quantity measurements on a construction site
     -when measuring less important detail
    g)-when measuring control lines for any survey
     -when measuring key components in a structural layout.

2.3 Distance = 38.00 - 0.19 = 37.81 ft.

2.4 H = 24.222 Cos 2°42' = 24.195 m

2.5 Tan slope angle = .02; angle = 1.14576°
     H = 179.36 Cos 1.14576° = 179.32 ft.

2.6 H = $\sqrt{91.333^2 - 0.75^2}$ = 91.33 m

2.7 Clearance = 32.7 + 2.1 = 34.8 ft.

2.8 Error per tapelength = -0.03 ft.; tape used 350.17/100 times
     error = -.03 x 3.5017 = -0.11 ft.
     corrected distance = 350.17 - 0.11 = 350.06 ft.

2.9 Error per tapelength = +0.006 m; tape used 119.898/30 times,
     error = +0.006 x 119.898/30 = +0.024 m
     corrected distance = 119.898 + 0.024 = 119.922 m

2.10 Error per tapelength = 0.04 ft.
Tape used a) 60/100 x .04 = +0.024 ft.
b) 100/100 x .04 = +0.040 ft.
Corrected layout distance a) 60.00 - 0.024 = 59.98 ft.
b) 100.00 - 0.040 = 99.96 ft.
(ie., these are the distances, which when pre-corrected for errors, will result in the required layout distances)

2.11 C_T = .00000645(100 - 68)210.50 = + 0.043 ft.
corrected distance = 210.50 + 0.04 = 210.54 ft.

2.12 C_T = .00000645(-10 -68)351.26 = - 0.177 ft.
C_L = -0.02 x 3.5126 = - 0.070 ft.

C = - 0.177 - 0.070 = - 0.25 ft.
Corrected slope distance = 351.26 -0.25 = 351.01 ft.
H = $\sqrt{351.01^2 - 4.62^2}$ = 350.98 ft.

2.13 C_T = .00000645(25 - 68)191.39 = - 0.05 ft.
Corrected slope distance = 191.39 - 0.05 = 119.34 ft.
H = 191.34 cos2°20' = 191.18 ft.

2.14 C_T = .0000116(25 - 20)211.416 = + 0.012m
C_L = -0.010 x 211.416/30 = - 0.070m
C = +0.012 - 0.070 = - 0.058 m
Corrected slope distance = 211.416 - 0.058 = 211.358m
H = 211.358 cos 3°42' = 210.917m

2.15 C_T = .0000116(0 - 20)300.000 = - 0.070m
C_L = +0.04 x 300/30 = + 0.040m

C = -0.070 +0.040 = - 0.030 m
Corrected slope distance = 300.000 - 0.030 = 299.970m
Tan slope angle = .015; slope angle = 0.8593722°
H = 299.970 cos 0.8593722° = 299.936m.

2.16 C_T = .00000645(100 -68)492.76 = + 0.10 ft.
C_L = +0.03 x 4.9276 = + 0.15 ft.
C = +0.10 +0.15 = + 0.25 ft.
Corrected slope distance = 492.76 + 0.25 = 493.01 ft.
Tan slope angle = .008; slope angle = 0.458356°
H = 493.01 cos 0.458356° = 492.99 ft.

2.17 C_T = .00000645(55-68)272.00 = - 0.023
C_L = .02 x 2.72 = - 0.054
C = -0.023 -0.054 = - 0.077 ft.= -0.08 ft.
Layout 272.00 + 0.08 = 272.08 ft.
(ie., this is the value, which when corrected by - 0.08 ft.,
will give the required distance of 272.00 ft.)

2.18 C_T = .0000116(22-20)242.666 = + 0.006 m
C_L = +0.012 x 242.666/30 = + 0.097
C = +0.006 +0.097 = + 0.103m
Layout 242.666 - 0.103 = 242.563 m

2.19 C_T = .0000116(15 -20)300.000 = - 0.017
C_L = -0.010 x 300/30 = - 0.100
C = -0.017 - 0.100 = - 0.117m
Layout 300.000 + 0.117 = 300.117m

2.20 C_T = .00000645(25-68)400.00 = - 0.111 ft.
C_L = +0.02 x 4.00 = +0.08 ft.
C = - 0.11 + 0.08 = -0.03 ft.
Layout 400.00 + .03 = 400.03 ft.

2.21 C_T = .00000645(100 - 68)420.00 = + 0.087
C_L = +.04 x 4.20 = + 0.168
C = +0.087 + 0.168 = + 0.255 ft.
Layout 420.00 -0.25 = 419.75 ft.

2.22 C_S = $-w^2L^3/24p^2$ = $-0.32^2 \times 48.888^3/(24 \times 100^2)$ = -.050m
Corrected distance = 48.888 -.050 = 48.838 m

2.23 30 m tape weighs 0.544 x 9.807 = 5.34 Newtons
C_S = $-W^2L/24p^2$ = $-5.34^2 \times 30/(24 \times 80^2)$ = -0.006m

2.24 C_S = $-W^2L/24p^2$ = -1.8^2 x 471.16/(24 x 24^2) = -0.11 ft.
Corrected distance = 471.16 -.11 = 471.05 ft.

2.25 C_S = $w^2L^3/24p^2$ = -0.016^2 x 72.55^3/(24 x 15^2) = -0.018 ft.
Corrected distance = 72.55 - 0.02 = 72.53 ft.

CHAPTER 3

3.1 a) (c+r) = 0.0675 x $(100/1000)^2$ = 0.0007m
    b) (c+r) = 0.0675 x $(200/1000)^2$ = 0.003m
    c) (c+r) = 0.574 x 0.5^2 = 0.14 ft.
    d) (c+r) = 0.574 x 5^2 = 14.35 ft.
    e) (c+r) = 0.0206 x $(800/1000)^2$ = 0.01 ft.
    f) (c+r) = 0.0206 x 4^2 = 0.33 ft.

3.2 5.5 = .574 K_1^2, K_1 = $\sqrt{5.5/.574}$ = 3.1 miles
    120 = .574 K_2^2, K_2 = $\sqrt{120/.574}$ = 14.5 miles

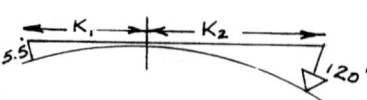

Maximum visibility distance = 17.6 miles

3.3 a) i   1.90   b) i   1.097
       ii   1.73   ii   1.055
       iii 1.57   iii 0.987
       iv   1.21   iv   0.950
       v   1.04   v   0.910

3.4

STATION	BS	HI	IS	FS	ELEVATION
BM #50	1.27	312.33			311.06
TP #1	2.33	309.75		4.91	307.42
TP #2				6.17	303.58

BS = 3.60   FS = 11.08
311.06 + 3.60 = 314.66 - 11.08 = 303.58 check

3.5

STATION	BS	HI	IS	FS	ELEVATION
BM #61	4.72	474.38			469.66
0+00			4.42		469.96
0+50			4.30		470.08
TP #1	5.11	477.48		2.01	472.37
1+00			4.66		472.82
1+50			3.98		473.50
1+75			1.20		476.28
TP #2				1.80	475.68 E = 0.002

BS = 9.83   FS = 3.81
469.66 + 9.83 = 479.49 - 3.81 = 475.68 check

3.6

STATION	BS	HI	FS	ELEVATION
BM 100	2.71	201.03		198.32
TP 1	3.62	199.77	4.88	196.15
TP 2	3.51	199.31	3.97	195.80
TP 3	3.17	199.67	2.81	196.50
TP 4	1.47	199.52	1.62	198.05
BM 100			1.21	198.31

BS = 14.48   FS = 14.49

198.32 + 14.48 14.14.49 = 198.31, check

3.7 Error of closure = 0.01 ft.; for 1000 ft., second order permits .035 1000/5280 = 0.015; therefore results qualify for **second order** accuracy.

3.8

STATION	BS	HI	IS	FS	ELEVATION
BM 20	8.27	396.52			388.25
TP 1	9.21	403.13		2.60	393.92
0+00			11.3		391.8
0+50			9.6		393.5
0+61.48			8.71		394.42
1+00			6.1		397.0
TP 2	7.33	405.80		4.66	398.47
1+50			5.8		400.0
2+00			4.97		400.83
BM 21				3.88	401.92

BS = 24.81   FS = 11.14
388.25 + 24.81 - 11.14 = 401.92   Check

3.9

STATION	BS	HI	IS	FS	ELEVATION
BM S101	0.475	168.747			168.272
0+000			0.02		168.73
0+020			0.41		168.34
0+040			0.73		168.02
0+060			0.70		168.05
0+066.28			.726		168.021
0+080			1.38		167.37
0+100			1.75		167.00
0+120			2.47		166.28
TP 1	0.666	166.420		2.993	165.754
0+140			0.57		165.85
0+143.78			0.634		165.786
0+147.02			0.681		165.739

0+160			0.71		165.71
0+180			0.69		165.73
0+200			1.37		165.05
TP 2	0.033	164.748		1.705	164.715
BM S102				2.891	161.857

BS = 1.174   FS = 7.589

168.272 + 1.174 - 7.589 = 161.875 check

3.10

STATION	BS	HI	IS	FS	ELEVATION
BM 21	1.203	242.212			241.009
0+00					
CL			1.211		241.001
10M LT., SL			1.430		240.782
10M RT., SL			1.006		241.206
0+20					
10M LT., SL			2.93		239.28
7.3M LT.			2.53		239.68
4M LT.			2.301		239.911
CL			2.381		239.831
4M RT.			2.307		239.905
7.8M RT.			2.41		239.80
10M RT., SL			2.78		239.43
0+40					
10M LT., SL			3.98		238.23
6.2M LT.			3.50		238.71
4M LT.			3.103		239.109
CL			3.187		239.025
4M RT.			3.100		239.112
6.8M RT.			3.37		238.84
10M RT., SL			3.87		238.34
TP 1				2.773	239.439

3.11

STATION	BS	HI	IS	FS	ELEVATION
BM 41	4.11	365.88			361.77
TP 13	4.10	369.09		0.89	364.99
12+00					
50 FT LT			3.9		365.2
18.3 FT LT			4.6		364.5
CL			6.33		362.76
20.1 FT RT			7.9		361.2

50 FT RT			8.2		360.9
13+00					
50 FT LT			5.0		364.1
19.6 FT LT			5.7		363.4
CL			7.54		361.55
20.7 FT RT			7.9		361.2
50 FT RT			8.4		360.7
TP 14	7.39	375.36	1.12		367.97
BM S.22			2.41		372.95

BS = 15.60   FS = 4.42
361.77 + 15.60 - 4.42 = 372.95 check

3.12

STATION	BS	HI	FS	ELEV.	LEFT		CL	RIGHT	
BM 37	7.20	518.40		511.20					
5+50					50 4.6 513.8	26.7 3.8 514.6	3.7 514.7	28.4 3.0 515.4	50 2.7 515.7
6+00					50 4.0 514.4	24.1 4.2 514.2	3.1 515.3	25.0 2.7 515.7	50 2.9 515.5
6+50					50 3.8 514.6	26.4 3.7 514.7	2.6 515.8	23.8 1.7 516.7	50 1.1 517.3
TP 1			6.71	511.69					

3.13

STATION	BS	HI	FS	ELEV.	LEFT		CL	RIGHT	
BM 107	7.71	388.82		381.11					
80+50					60 9.7 379.1	28 8.0 380.8	5.7 383.1	32 4.3 384.5	60 4.0 384.8
81+00					60 10.1 378.7	25 9.7 379.1	6.8 382.0	30 6.0 382.8	60 5.3 383.5
81+50					60 11.7 377.1	27 11.0 377.8	9.2 379.6	33 8.3 380.5	60 8.0 380.8
TP 1			10.17	378.65					

3.14 a) V = 148.61 Sin 21°26' = 54.30 ft
Elevation of lower station = 829.76 +4.66 -54.30 -4.88 = 775.24 ft.
b) H = 148.61 Cos 21° 26' = 138.33 ft
Lower station at 110+71.25 +138.33 = 112+09.58

3.15

STATION	BS	HI	FS	ELEVATION
BM 130	0.702	171.928		171.226
TP 1	0.970	171.787	1.111	170.817
TP 2	0.559	171.667	0.679	171.108
TP 3	1.744	170.631	2.780	168.887
BM K110	1.973	170.936	1.668	168.963
TP 4	1.927	171.075	1.788	169.148
BM 132			0.888	170.187

BS = 7.875   FS = 8.914
171.226 +7.875 -8.914 = 170.187, check

3.15 a) Error = 170.198-170.187 = -0.011m
Accuracy limit for 2nd order   = .007 $\sqrt{.8}$ = .006
Accuracy limit for 3rd order   = .012 $\sqrt{.8}$ = .011 (U.S.)
            or   = .024 $\sqrt{.8}$ = .021 (Canada)
(SEE TABLES A.18 & A.19)
Therefore the error of -0.011 satisfies the requirements for 3rd order accuracy in both the U.S. and Canada.

3.15 b)

STATION	CUMULATIVE DISTANCE	ELEVATION	CORRECTION	ADJUSTED ELEVATION
BM 130		171.226		171.226
TP 1	130	170.817	130/780 x.011 = +.002	170.819
TP 2	260	171.108	260/780 x.011 = +.004	171.112
TP 3	390	168.887	390/780 x.011 = +.006	168.893
BM K110	520	168.963	520/780 x.011 = +.007	168.970
TP 4	650	169.148	650/780 x.011 = +.009	169.157
BM 132	780	170.187	780/780 x.011 = +.011	170.198

The adjusted elevation of BM 132 is 170.198m

3.16 a) True difference = 6.27-3.78 = 2.49 ft.
b) Correct rod reading = 5.21+2.49 = 7.70 ft. on A
c) Error is +0.06 in 260 ft.,or .0002 ft/ft
d) Cross hair adjusted downward to read 7.70 on A

3.17 a) 422.38+2.86-10.77 = 414.47 ft.(TBM)
net difference between BS and FS = 150 ft.
error in 150 ft. = .0002 x 150 = 0.03 ft.(too high)
422.38+2.86-10.80 = 414.44 ft.(TBM)
b) (c+r) error over a net distance of 150 ft. is negligible, ie., (c+r) = .0206 $(.15)^2$ = .0005 ft.

3.18 a) First elevation difference = 2.417-0.673 = 1.744
Second elevation difference = 2.992-1.252 = 1.740
Average elevation difference = 1.742
Elevation B = 187.298-1.742 = 185.556
b) The leveling error is 0.004m

3.19 AB: 2000.00 Sin 3°30' + .0206$(2)^2$ = +122.18'
AC: 2000.00 Sin 1°30' - .0206 $(2)^2$ = -52.27'
AD: (c+r) = .0206$(3)^2$ = +0.19'

3.20 Elev. Difference: 6301.76 Sin$(2°45'30")^2$ = 303.26'
(c+r) + .0206(6.3017 Cos 2°45'30"$)^2$ = +0.82
630.15 + 304.08 = 934.25'

3.21 a) 1879.2195 Cos1°26'44" = 1878.61 ft.
b) 150.166 + 1879.2195 Sin 1°26'44" = 197.57 ft.

CHAPTER 4

4.1
(n-2)180 = 3 X 180 = 540°00'
A = 110°27'00"
B = 130°52'30"
C = 88°17'00"   539°60'
D = 97°33'30"   -426°10'
E = ?   E = 113°50'
426°10'00"

4.2
a) S19°37'W
b) S51°09'E
c) S10°57'E
d) N84°29'W

e) N17°17'E
f) N59°40'W
g) S26°51'W

4.3
a) 66°27'
b) 349°50'
c) 170°54'
d) 261°28'

e) 89°59'
f) 185°08'
g) 149°30'

4.4
a) 19°37'
b) 308°51'
c) 349°03'
d) 95°31'

e) 197°19'
f) 120°20'
g) 26°11'

4.5
a) S66°27'W
b) S10°10'E
c) N 9°06'W
d) N81°28'E

e) S89°59'W
f) N 5°08'E
g) N30°30'W

4.6
AB N40°37'E
+B   8°13'
BC N48°50'E
+C   2°21'
CD N51°11'E
+D   14°41'
DE N65°52'E

DE N65°52'E
-E   21°08'
EF N44°44'E
-F   6°32'
FG N38°12'E
+G   1°15'
GH N39°27'E

10

4.7
A   61°27'
+ 63°41'
   125°08'

   179°60'
- 125°08'
A = 54°52'

B   61°27'
+ 48°31'
B = 109°58'

C   48°31'
+ 16°20'
   64°51'

   179°60'
-   64°51'
C = 115°09'

D   16°20'
+ 63°41'
D = 80°01'

GEOMETRIC CHECK
A 54°52'
B 109°58'
C 115°09'
D 80°01'
   358°120'
 = 360°00'
CHECK

4.8 <u>CLOCKWISE SOLUTION</u>
       B =   140°28'50"
   -BRG BA   36°18'20"
         =   104°10'30"

                179°59'60"
               -104°10'30"
    BRG BC =   N 75°49'30"E

        C =   101°30'20"
   -BRG CB   75°49'30"
    BRG CD =   S 25°40'50"E

        D =   72°48'10"
   +BRG DC   25°40'50"
                98°29'00"
    BRG DE =   S 81°31'00"W

11

```
        E =    161°25'40"
   -BRG ED     81°31'00"
    BRG EA =  N 79°54'40"W

        A =    63°47'00"    179°59'60"
   +BRG AE     79°54'40"   -143°41'40"
          =   143°41'40"   N36°18'20"E (BRG AB)
```

COUNTER-CLOCKWISE SOLUTION

```
        A =    63°47'00"
   +BRG AB     36°18'20"
          =   100°05'20"

                179°59'60"
              - 100°05'20"
    BRG AE =   S79°54'40"E

        E =    161°25'40"
   -BRG EA     79°54'40"
    BRG ED =   N81°31'00"E


   BRG DE =    81°31'00"
      + D      72°48'10"
               154°19'10"

                179°59'60"
              - 154°19'10"
    BRG DC =   N25°40'50"W

        C =    101°30'20"
   -BRG CD     25°40'50"
    BRG CB =   S75°49'30"W

        B =    140°28'50"
   +BRG BC     75°49'30"
          =   216°18'20"
              -180°
    BRG BA=   S36°18'20"E  CHECK
```

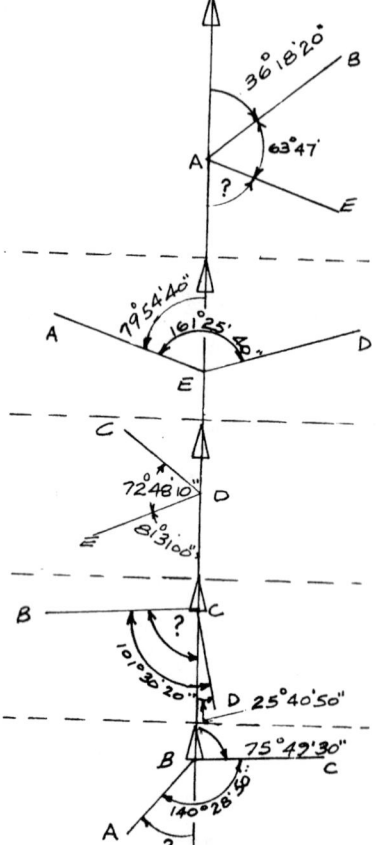

12

4.9 CLOCKWISE SOLUTION

Azimuth AB =	36°18'20"
+	180
Azimuth BA =	216°18'20"
-B	140°28'50"
Azimuth BC =	75°49'30"
+	180
Azimuth CB =	255°49'30"
-C	101°30'20"
Azimuth CD =	154°19'10"
+	180
Azimuth DC =	334°19'10"
-D	72°48'10"
Azimuth DE =	261°31'00"
+	180
Azimuth ED =	441°31'00"
-E	161°25'40"
Azimuth EA =	280°05'20"
-	180
Azimuth AE =	100°05'20"
-A	63°47'00"
Azimuth AB =	36°18'20" **CHECK**

COUNTER-CLOCKWISE SOLUTION

Azimuth AB =	36°18'20"
+A	63°47'00"
Azimuth AE =	100°05'20"
+	180
Azimuth EA =	280°05'20"
+E	161°25'40"
Azimuth ED =	441°31'00"
-	360
Azimuth ED =	81°31'00"
+	180
Azimuth DE =	261°31'00"
+D	72°48'10"
Azimuth DC =	334°19'10"
-	180
Azimuth CD =	154°19'10"
+C	101°30'20"
Azimuth CB =	255°49'30"
-	180

Azimuth BC = 75°49'30"
         +B 140°28'50"
Azimuth BA = 216°18'20"
          -    180
Azimuth AB = 36°18'20" **CHECK**

4.10
B = 140°28'50"
-BRG B 52°13'00"
BRG BC = S88°15'10"E

C = 101°30'20"
+BRG CB 88°15'10"
         189°46'10"
       - 180°
BRG CD = S 9°46'10"E

D = 72°48'10"
+BRG DC 9°46'10"
BRG DE = N82°34'20"W

E = 161°25'40"
+BRG ED 82°34'20"
         244°00'00"
       - 180°
BRG EA = N64°00'00"W

A = 63°47'00"
+BRG AE = 64°00'00"
       = 127°47'00"
         179°60'
       - 127°47'
BRG AB = N52°13'00"E **CHECK**

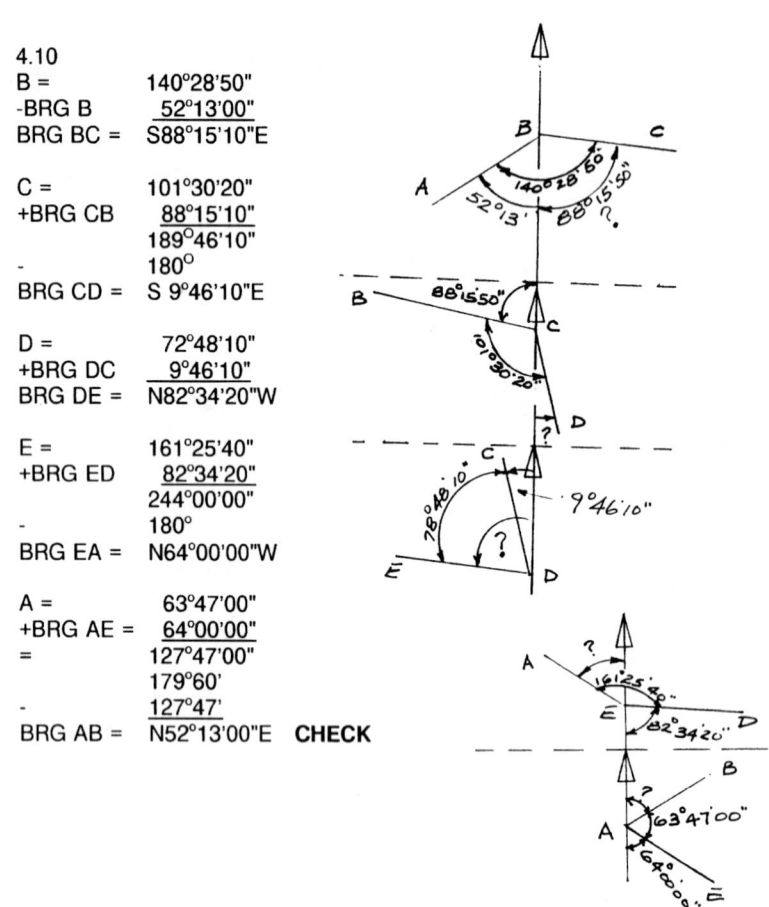

14

4.11

Azimuth AB =	52°13'00"
+A	63°47'00"
Azimuth AE =	116°00'00"
+	180
Azimuth EA =	296°00'00"
+E	161°25'40"
Azimuth ED =	457°25'40"
-	360
Azimuth ED =	97°25'40"
+	180
Azimuth DE =	277°25'40"
+D	72°48'10"
Azimuth DC =	350°13'50"
-	180
Azimuth CD =	170°13'50"
+C	101°30'20"
Azimuth CB =	271°44'10"
-	180
Azimuth BC =	91°44'10"
+B	140°28'50"
Azimuth BA =	232°13'00"
-	180
Azimuth AB =	52°13'00" **CHECK**

4.12 Using Figures 4.14(b) & (c)
a) On Jan. 2/90, at Seattle, Wash., the declination [scaled from Figure 4.14 c] is 20°18' - 0°28'= 19°50' [5yrs x 05.6'-annual change also scaled from Figure 4.14 c] .

Annual change = 0°05.6' x 21.5 years = 2°00'
Magnetic bearing on July 2,1968 = N39°30'E - 2°00' = N37°30'E

b) Astronomic bearing = N39°30'E + 19°50' = N59°20'E

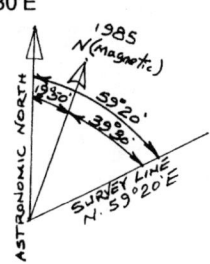

CHAPTER 6

6.1

STATION	FIELD ANGLES	CORRECTIONS	ADJUSTED ANGLES
A	88°10'00"	+ 30"	88°10'30"
B	124°38'00"	+ 30"	124°38'30"
C	95°57'30"	+ 30"	95°58'00"
D	109°36'30"	+ 30"	109°37'00"
E	121°35'30"	+ 30"	121°36'00"
	537°176'90"	150"	537°179'60"
	539°57'30"	=	540°00'00"

Error = 2'30", = 150"; Correction/angle = 30"

6.2a

STATION	FIELD ANGLES	CORRECTIONS	ADJUSTED ANGLES
A	81°22'30"	+ 30"	81°23'
B	72°32'30"	+ 30"	72°33'
C	89°39'30"	+ 30"	89°40'
D	116°23'30"	+ 30"	116°24'
	358°116'120"	120"	358°120'
	359°58'		360°00'

Error = 02', or 120"; Correction/angle = 30"

6.2b

Course	Azimuth	Bearing	Distance	Latitude	Departure
AB	205°25'	S25°25'W	636.45	-574.85	-273.16
BC	97°58'	S82°02'E	654.45	- 90.70	648.13
CD	7°38'	N 7°38'E	382.85	379.46	50.86
DA	304°02'	N55°58'W	512.77	286 .98	-424.94
			2186.56	+ 0.89	+ 0.89

$E = \sqrt{.89^2 + 0.89^2} = 1.26$ ft.
Accuracy = E/P = 1.26/2186.56 = 1/1737 = 1/1700

6.3a Corrections

Course	C_{Lat}	C_{Dep}	Adjusted Lat.	Adjusted Dep.
AB	-0.26	-0.26	-575.11	-273.42
BC	-0.27	-0.27	- 90.97	+647.86
CD	-0.16	-0.16	+379.30	+ 50.70
DA	-0.20	-0.20	+286.78	-425.14
	-0.89	-0.89	0.00	0.00

6.3b Coordinates

STATION	NORTH	EAST
B	1000.00	1000.00
	-90.97	+647.86
C	909.03	1647.86
	+379.30	+50.70
D	1288.33	1698.56
	+286.78	-425.14
A	1575.11	1273.42
	-575.11	-273.42
B	1000.00	1000.00

6.4a Area by coordinates

$X_B(Y_C-Y_A)$ = 1000.00(909.03-1575.11) = - 666,080
$X_C(Y_D-Y_B)$ = 1647.86(1288.33-1000.00) = + 475,127
$X_D(Y_A-Y_C)$ = 1698.56(1575.11 - 909.03) = +1,131,377
$X_A(Y_B-Y_D)$ = 1273.42(1000.00-1288.33) = - 367,165
                                2A =   573,259
                                 A = 286,629 ft^2
(1 Acre = 43,560 ft^2)   A = 6.58 Acres

6.4b Area by DMD's

COURSE	DMD	LATITUDE	DOUBLE AREA
BC	+647.86	-90.97	- 58,936
	+647.86		
	+50.70		
CD	+1346.42	+379.30	+510,697
	+50.70		
	-425.14		
DA	+971.98	+286.78	+278,744
	-425.14		
	-273.42		
AB	+273.42	-575.11	-157,247

                                2A =   573,258
                                 A = 286,629 ft^2
(1 Acre = 43,560 ft^2)   A = 6.58 Acres

6.5

	ANGLES
A	101°28'26"
B	102°10'42"
C	104°42'06"
D	113°04'42"
E	118°34'04"

538°118'120"
540°00'00" CHECK

6.5 a,b

CRSE	AZIMUTH	BEARING	DISTANCE	LATITUDE	DEPARTURE
AE	182°20'29"	S 2°20'29"W	20.845	-20.828	- 0.852
ED	120°54'33"	S59°05'27"E	35.292	-18.129	+30.280
DC	53°59'15"	N53°59'15"E	37.090	+21.808	+30.002
CB	338°41'21"	N21°18'39"W	26.947	+25.104	- 9.793
BA	260°52'03"	S80°52'03"W	50.276	- 7.980	- 49.639
			170.450	- 0.025	- 0.002

6.5c $E = \sqrt{0.025^2 + 0.002^2} = 0.025$ m
Accuracy = E/P = 0.025/170.450 = 1/6818 = 1/6800

6.6A Balance using Compass Rule

COURSE	C_{LAT}	C_{DEP}	Corrected LAT	Corrected DEP
AE	+.003	+.000	-20.825	- 0.852
ED	+.005	+.000	-18.124	+30.280
DC	+.006	+.001	+21.814	+30.003
CB	+.004	+.000	+25.108	- 9.793
BA	+.007	+.001	- 7.973	-49.638
CHECK -	+.025	+.002	0.000	0.000

6.6b COORDINATES

STATION	NORTH	EAST
A	1000.000	1000.000
	- 20.825	+ 0.852
E	979.175	999.148
	- 18.124	+ 30.280
D	961.051	1029.428
	+ 21.814	+ 30.003
C	982.865	1059.431
	+ 25.108	- 9.793
B	1007.973	1049.638
	- 7.973	- 49.638
A	1000.000	1000.000 **CHECK**

6.7a Area by the Coordinate Method
$X_A(Y_B-Y_E)$ = 1000.000(1007.973-979.175) = +28,798
$X_E(Y_A-Y_D)$ = 999.148(1000.000-961.051) = +38,916
$X_D(Y_E-Y_C)$ = 1029.428(979.175-982.865) = - 3,799
$X_C(Y_D-Y_B)$ = 1059.431(961.051-1007.973) = -49,711
$X_B(Y_C-Y_A)$ = 1049.638(982.865-1000.000) = -17,986
$$2A = 3{,}781 \text{ m}^2$$
$$\text{Area, } A = 1{,}890 \text{ m}^2$$

6.7b Area by DMD's

COURSE	DMD	LATITUDE	DOUBLE AREA
ED	30.280	-18,124	- 549
	30.280		
	30.003		
DC	90.563	+21.814	+1,976
	30.003		
	-9.793		
CB	110.773	+25.108	+2,781
	- 9.93		
	-49.638		
BA	+51.342	- 7.973	- 409
	-49.638		
	- 0.852		
AE	+ 0.852	-20.825	- 18

$$2A = 3781 \text{ m}^2$$
$$\text{Area, } A = 1890 \text{ m}^2$$

6.8

COURSE	BEARING	DISTANCE	LATITUDE	DEPARTURE
AB	N70°10'07"E	80.32	+ 27.25	+ 75.56
BC	N74°29'00"E	953.83	+255.17	+919.07
CD	N70°22'45"E	818.49	+274.84	+770.96
AD			+557.26	+1765.59

Distance AD = $\sqrt{557.26^2 + 1765.59^2}$ = 1851.44 ft.

Tan Bearing AD = 1765.59/557.26; Brg. AD = N72°28'59"E

6.9

COURSE	BEARING	DISTANCE	LATITUDE	DEPARTURE
EA	N26°58'31"W	483.669	+431.047	-219.395
AB	N37°10'49"E	537.144	+427.963	+324.609
BC	N79°29'49"E	1109.301	+202.212	+1090.715
EC			+1061.222	+1195.929

Distance EC = $\sqrt{1061.222^2 + 1195.929^2}$ = 1598.887m

Tan Bearing EC = 1195.929/1061.222; Brg EC = N48°24'55"E

Angle @ C = 48°24'55"
             - 18°56'31"

           = 29°28'24"

Sin D = Sin 29°28'24"/953.829 x 1598.887
D = 55°33'52" or 124°26'08"
From the given data, D = 124°26'08"

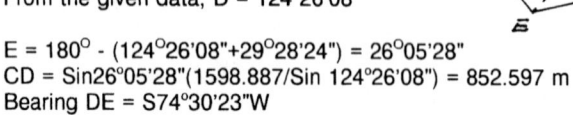

E = 180° - (124°26'08"+29°28'24") = 26°05'28"
CD = Sin26°05'28"(1598.887/Sin 124°26'08") = 852.597 m
Bearing DE = S74°30'23"W

6.10
AB = $\sqrt{(738.562-559.319)^2+(666.737-207.453)^2}$ = 493.021m
Tan Brg AB = 459.284/179.243; Brg. AB = N68°40'52"E
BC = $\sqrt{(541.742-738.562)^2+(688.350-666.737)^2}$ = 198.003m
Tan Brg BC = 21.613/-196.820; Brg. BC = S 6°16'00"E
CD = $\sqrt{(379.861-541.742)^2+(839.008-688.350)^2}$ = 221.141m
Tan Brg CD = 150.658/-161.881; Brg. CD = S42°56'36"E

$DE = \sqrt{(296.099-379.861)^2+(604.048-839.008)^2} = 249.444m$
Tan Brg DE = -234.960/-83.762; Brg. DE = S70°22'45"W
$EF = \sqrt{(218.330-296.099)^2+(323.936-604.048)^2} = 290.707m$
Tan Brg EF = -280.112/-77.769; Brg. EF = S74°29'00"W
$FA = \sqrt{(559.319-218.330)^2+(207.453-323.936)^2} = 360.336m$
Tan Brg FA = -116.483/340.989; Brg. FA = N18°51'37"W

6.11
$X_A(Y_B-Y_F)$ = 207.453(738.562-218.330) = +107,924
$X_B(Y_C-Y_A)$ = 666.737(541.742-559.319) = - 11,719
$X_C(Y_D-Y_B)$ = 688.350(379.861-738.562) = -246,912
$X_D(Y_E-Y_C)$ = 839.008(296.099-541.742) = -206,096
$X_E(Y_F-Y_D)$ = 604.048(218.330-379.861) = - 97,572
$X_F(Y_A-Y_E)$ = 323.936(559.319-296.099) = + 85,266

2A = 369,109m²
Area, A = 184,555m²
or, A = 18.455 hectares

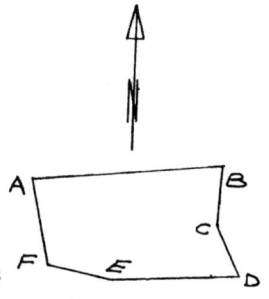

6.12
LINE AD
Y-559.319 = (379.861-559.319)/(839.008-207.453)x (X-207.453)
0.28415X + Y = 618.2673 **(1)**
LINE FB
Y-218.330 = (738.562-218.330)/(666.737-323.936)x (X-323.936)
1.51759X - Y = 273.2726 **(2)**

1.80174X = 891.5399   **(1) + (2)**
X = 494.8204
Y = 477.6628
Coordinates of K = 477.633N, 494.820E

LINE EB
Y-296.099 = (738.562-296.099)/(666.737-604.048)x (X-604.048)
7.05806X - Y = 3967.3106 **(3)**
LINE AC
Y-559.319 = (541.742-559.319)/(688.350-207.453)x (X-207.453)
0.03655X + Y = 566.9015 **(4)**

7.0946X = 4534.212   **(3) + (4)**
X = 639.106
Y = 543.542
Coordinates of L = 543.542N, 639.106E

21

Tan Brg KL = (639.106-494.820)/(543.542-477.663);
Brg KL = N65°27'33"E
KL = $\sqrt{144.286^2 + 65.879^2}$ = 158.614m

6.13

COURSE	AZIMUTH	BEARING	DISTANCE	LATITUDE	DEPARTURE
AE	159°39'40"	S20°20'20"E	20.845	-19.545	+ 7.245
ED	98°13'44"	S81°46'16"E	35.292	- 5.051	+34.929
DC	31°18'26"	N31°18'26"E	37.090	+31.689	+19.273
CB	316°00'32"	N43°59'28"W	26.947	+19.387	-18.716
BA	238°11'14"	S58°11'14"W	50.276	-26.503	-42.723
			170.450	- 0.023	+ 0.008

E = $\sqrt{0.023^2 + 0.008^2}$ = 0.0244m
Accuracy = E/P = 0.0244/170.450 = 1/7000

6.14

COURSE	C_{LAT}	C_{DEP}	Corrected LAT.	Corrected DEP.
AE	+0.003	-0.001	-19.542	+ 7.244
ED	+0.005	-0.002	- 5.046	+34.927
DC	+0.005	-0.002	+31.694	+19.271
CB	+0.004	-0.001	+19.391	- 18.717
BA	+0.006	-0.002	-26.497	- 42.725
CHECK	+0.023	-0.008	0.000	0.000

6.15

STATION	COORDINATES	
	NORTH	EAST
A	1000.000	1000.000
E	- 19.542 980.458	+ 7.244 1007.244
D	- 5.046 975.412	+ 34.927 1042.171
C	+ 31.694 1007.106	+ 19.271 1061.442
B	+ 19.391 1026.497	- 18.717 1042.725
A	- 26.497 1000.000	- 42.725 1000.000 CHECK

6.16 Area by Coordinates

$X_A(Y_B-Y_E) = 1000.000(1026.497-980.458) = +46,039$
$X_E(Y_A-Y_D) = 1007.244(1000.000-975.412) = +24,766$
$X_D(Y_E-Y_C) = 1042.171(980.458-1007.106) = -27,772$
$X_C(Y_D-Y_B) = 1061.442(975.412-1026.497) = -54,224$
$X_B(Y_C-Y_A) = 1042.725(1007.106-1000.000) = +7,410$

$2A = 3,781 \text{ m}^2$
AREA, $A = 1,890 \text{ m}^2$

Area by DMD'S

COURSE	DMD	LATITUDE	DOUBLE AREA
AE	7.244 +7.244 +34.927	-19.542	-142
ED	49.415 +49.415 +19.271	-5.046	-249
DC	103.613 +19.271 -18.717	+31.694	+3,284
CB	104.167 -18.717 -42.725	+19.391	+2,020
BA	42.725	-26.497	-1,132

$2A = 3,781 \text{ m}^2$
$A = 1,890 \text{ m}^2$

6.17 COORDINATES

STATION	NORTH	EAST
K	1990.000	2033.000
	+ 10.000	- 33.000
A	2000.000	2000.000
K	1990.000	2033.000
	+ 25.271	+ 19.455
B	2015.271	2052.455
K	1990.000	2033.000
	- 0.311	+ 38.285
C	1989.689	2071.285
K	1990.000	2033.000
	- 30.055	+ 7.245
D	1959.945	2040.245
K	1990.000	2033.000
	- 12.481	- 30.100
E	1977.519	2002.900

6.18
$X_A(Y_E-Y_B) = 2000.000(1977.519-2015.271) = -75,504$
$X_B(Y_A-Y_C) = 2052.455(2000.000-1989.689) = +21,163$
$X_C(Y_B-Y_D) = 2071.285(2015.271-1958.945) = +114,596$
$X_D(Y_C-Y_E) = 2040.245(1989.689-1977.519) = +24,830$
$X_E(Y_D-Y_A) = 2002.900(1959.945-2000.000) = -80,226$
$$2A = 4,859 \text{ m}^2$$
$$\text{AREA, A} = 2,430 \text{ m}^2$$

6.19 COURSE	BEARING	DISTANCE (m)
AB	N73°46'06"E	54.633
BC	S36°21'20"E	31.765
CD	S46°13'17"W	42.991
DE	N64°47'56"W	41.273
EA	N 7°20'52"W	22.675

CHAPTER 7

7.1

STA.	ROD INTERVAL	VERTICAL ANGLE	HORIZONTAL DISTANCE	ELEVATION DIFFERENCE	ELEVATION
	⊼ Station				409.58
1	3.48	+0°58'	347.9	+ 5.9	415.5
2	0.38	-3°38'	37.8	- 2.4	407.2
3	1.40	-1°30'	139.9	- 3.7	415.9
4	2.49	+0°20'	249.0	+ 1.4	411.0
5	1.11	+2°41'	110.8	+ 5.2	414.8

7.2

STA.	ROD INT.	VERTICAL ANGLE	HORIZ. DIST.	ELEVATION DIFFERENCE	ELEVATION
		⊼ Station, hi = 1.83			143.78
1	0.041	+2°19'	4.09	+0.17	143.95
2	0.072	+1°57' on 1.43	7.19	V=+0.24	144.42
3	0.555	0°00' on 2.71	55.5	V=0	142.90
4	1.313	-2°13'	131.10	-5.07	138.71
5	1.111	-4°55' on 1.93	110.28	V=-9.49	134.19
6	0.316	+0°30'	31.6	+0.28	144.06

7.3

a) V=100 x 1.31 x Cos 3°12' Sin 3°12' = 7.30 ft.
   Elev.K + hi (5.36) + V (7.30) -RR (4.27) = ELEV.L (382.88)
   Elev.K = 374.49 ft.

b) H = 100 x 1.31 x Cos^{2}3°12' = 130.59 ft.

7.4

STA.	ROD INT.	HORIZ. ANGLE	VERTICAL ANGLE	HORIZ. DIST.	ELEV. DIFF.	ELEV.
		⊼ @ Mon. #36, hi = 5.14				532.33
#37	3.22	0°00'	+3°46'	320.6	+21.1	553.4
1	2.71	2°37'	+2°52'	270.3	+13.5	545.8
2	0.82	8°02'	+1°37'	81.9	+ 2.3	534.6
3	1.41	27°53'	+2°18'on 4.06	140.8	V=5.7	539.1
4	1.10	46°20'	+0°18'	110.0	+ 0.6	532.9
5	1.79	81°32'	0°00'on 8.11	179.0	V=0	529.4
6	2.61	101°17'	-1°38'	260.8	- 7.4	524.9
#38	3.60	120°20'	-3°41'	358.5	-23.1	509.2

7.5

STA.	ROD INT.	HORIZ. ANGLE	VERT. ANGLE	HORIZ. DIST.	ELEV. DIFF.	ELEV.
		⊼ @ K, hi = 1.82				141.66
L		0°00'				
0+00 CL	0.899	34°15'	-19°08'	80.24	-27.84	113.82
S Ditch	0.851	33°31'	-21°58'	73.19	-29.52	112.14
N Ditch	0.950	37°08'	-20°42'	83.13	-31.41	110.25
0+50 CL	0.622	68°17	-16°58'	56.90	-17.36	124.30
S Ditch	0.503	64°10'	-20°26'	44.17	-16.46	125.20
N Ditch	0.687	70°48'	-19°40'	60.92	-21.77	119.89
1+00 CL	0.607	113°07'	-13°50'	57.23	-14.09	127.57
S ditch	0.511	109°52'	-16°48'	46.83	-14.14	127.52
N ditch	0.710	116°14'	-14°46'	66.39	-17.50	124.16
1+50 CL	0.852	139°55'	-10°04'	82.60	-14.66	127.00
S ditch	0.800	135°11'	-11°22'	76.89	-15.46	126.20
N ditch	0.932	144°16'	-10°32'	90.09	-16.75	124.91
2+00 CL	1.228	152°18'	- 6°40'	121.14	-14.16	127.50

S ditch	1.148	155°43'	- 8°00'	112.58	-15.82	125.84
N ditch	1.263	147°00'	- 7°14'	124.30	-15.78	125.88

7.6

COURSE	HORIZ.DIST.	VERT.DIST.	STA.	ELEVATION
AB	68.10	+2.616	A	190.64
BA	68.50	-2.612		+ 2.61
MEAN	68.30	2.614	B	193.25
				+ 6.37
BC	82.10	+6.39	C	199.62
CB	82.21	-6.35		- 3.26
MEAN	82.155	6.37	D	196.36
				- 1.85
CD	73.15	-3.26	E	194.51
DC	73.13	+3.25		- 3.88
MEAN	73.14	3.255	A	190.63
			Error =	0.01m
DE	60.54	-1.83		
ED	60.54	+1.87		
MEAN	60.54	1.85		
EA	113.57	-3.87		
AE	113.57	+3.90		
MEAN	113.57	3.885		

CHAPTER 8

8.1 Prism constant = EG-EF-FG
= 586.645-298.717-287.958 = -0.030m

8.2 H = 5170.11 Cos 2°45'30" = 5164.12 ft.
V = 5170.11 Sin 2°45'30" = 248.80 ft.
Elevation of target station = 630.15+248.80 = 878.95 ft.

8.3 @ A, H = 1879.209 Cos 1°26'50" = 1878.610m
Elevation difference = 1879.209 Sin 1°26'50" = 47.462m
@ B, H = 1879.230 Cos 1°26'38" = 1878.633m
Elevation difference = 1879.230 Sin 1°26'38" = 47.353m
a) Horizontal distance = (1878.610+1878.633)/2 = 1878.622m
b) Elev. B = 181.302 + (47.462+47.353)/2 = 228.71m

8.4 LL^1 = 2000.00 Sin 3°30' =+122.10 ft.
c+r = .0206 x 2^2 = +0.08 ft.
Elevation difference, K to L, = +122.18 ft
Elevation of L = 241.69+122.18 = 363.87 ft.

MM^1 = 2000.00 Sin -1°30' = -52.35 ft.
c+r = .0206 x 2^2 = +0.08 ft.
Elevation difference, K to M, = -52.27 ft.
Elevation of M = 241.69-52.27 = 189.42 ft.

c+r = .0206 x 3^2 = + 0.19 ft.
Elevation difference, K to N, = +0.19 ft.
Elevation of N = 241.69+0.19 = 241.88 ft.

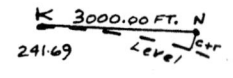

8.5 ΔHR - Δhi = 0.150 - 0.100 = 0.050m
Sin Δα = (0.050 Cos 4°18'30")/387.603
Δα = 0°00'27"
$α_K$ = 0°00'27"+4°18'30" = 4°18'57"
H = 387.603 Cos 4°18'57" = 386.504m
Elevation of B = 110.222+1.601+(387.603 Sin 4°18'57")-1.915 = 139.077m.

8.6 ΔHR-Δhi = 0.39-0.31 = 0.08 ft.
Sin Δα = (0.08 Cos 3°14'30")/536.88
Δα = 0°00'31"
$α_K$ = 0°00'31"+3°14'30" = 3°15'01"
H = 536.88 Cos 3°15'01" = 536.02 ft.
Elevation B = 531.49+5.21+(536.88 Sin 3°15'01")-5.78 = 561.36 ft.

CHAPTER 9
9.1

TRAVERSE COMPUTATIONS

COURSE	BEARING	DISTANCE	LATITUDE	DEPARTURE
AB	N. 3°30'E.	56.05	+ 55.95	+ 3.42
BC	N. 0°30'W.	61.92	+ 61.92	- 0.54
CD	N.88°40'E.	100.02	+ 2.33	+ 99.99
DE	S.23°30'E.	31.78	- 29.14	+ 12.67
EF	S.28°53'W.	69.11	- 60.51	+ 33.38
FG	SOUTH	39.73	- 39.73	0
GA	N.83°37'W.	82.67	+ 9.18	- 82.16
			0.00	0.00

9.2

INTERIOR ANGLES
A -   92°53'
B - 184°00'   CHECK
C -   90°50'   (n-2) 180,
D - 112°10'   5 x 180 = 900°
E - 127°37'
F - 208°53'
G -   83°37'
     896°240'
 =   900°00'

9.5d

AREA COMPUTATIONS

COURSE	D.M.D.	LATITUDE	2 X AREA
AB	+3.42	+ 55.95	+ 191
	+3.42		
	-0.54		
BC	6.30	+ 61.92	+ 390
	-0.54		
	+99.99		
CD	105.75	+ 2.33	+ 246
	+99.99		
	+12.67		
DE	218.41	- 29.14	- 6364
	+12.67		
	-33.38		
EF	197.70	- 60.51	- 11963
	-33.38		
	0		
FG	164.32	- 39.73	- 6528
	0		
	-82.16		
GA	82.16	+ 9.18	+ 754

2A = - 23,274
A =   11,637 m^2

or A = 11,637 ft^2
  = 0.27 acres

#9.8

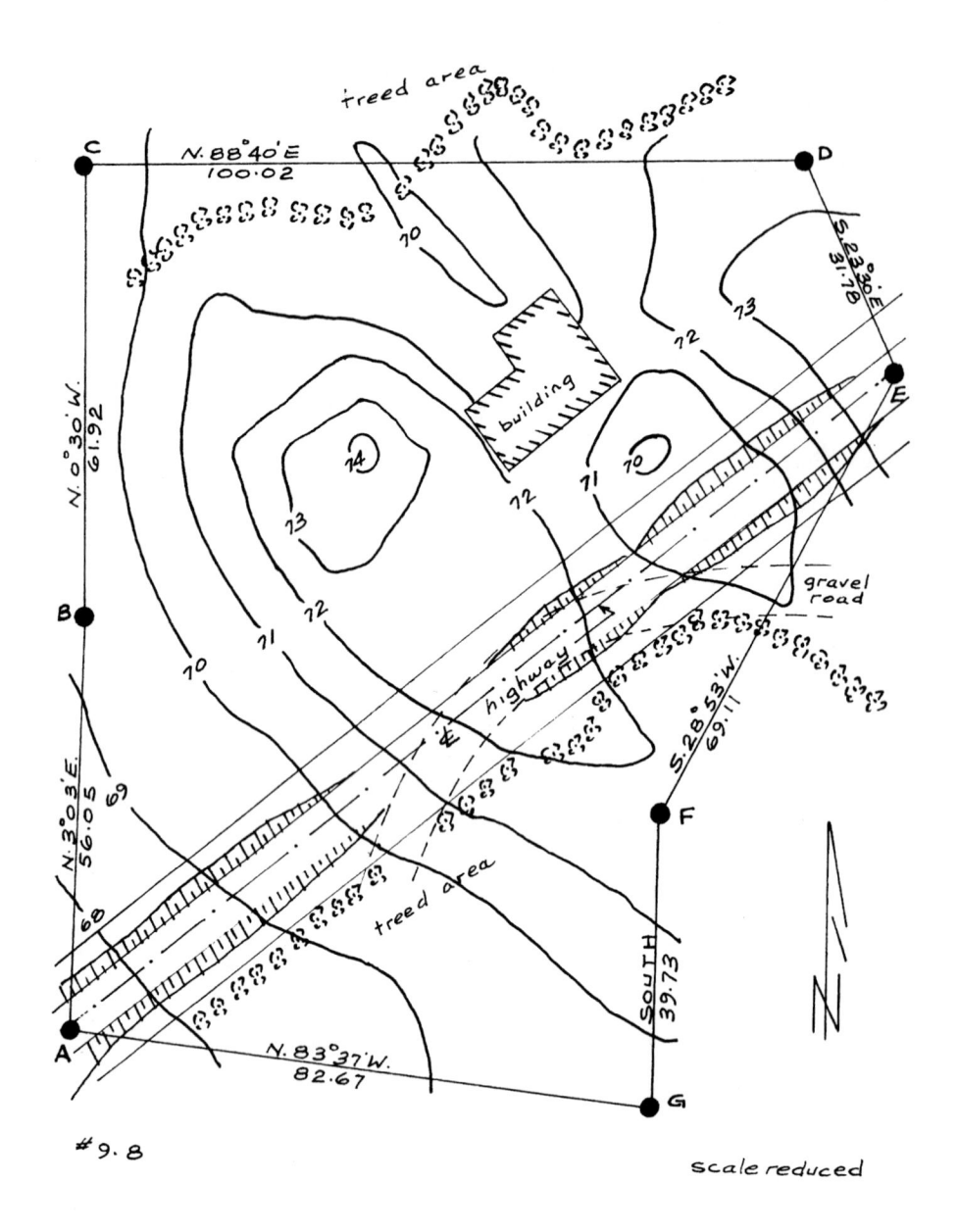

scale reduced

CHAPTER 10

10.1 T = RTan Δ/2 = 700Tan14°51' = 185.60 ft.
L = $\frac{2\pi R \Delta}{360}$ = $\frac{2\pi 700 \times 29.7}{360}$ = 362.85 ft.

10.2 R = $\frac{5729.58}{D}$ = $\frac{5729.58}{8}$ = 716.20 ft.

T = Rtan Δ/2 = 716.20 Tan3°35' = 44.85 ft.
L = $\frac{100\Delta}{D}$ = $\frac{100 \times 7.166}{8}$ = 89.58 ft.

10.3 PI @   9 + 27.26
     - T   1   85.60
     BC =   7 + 41.66
     +L   3   62.85
     EC =   11 + 04.51

10.4 PI @   15 + 88.10
     - T   44.85
     BC =   15 + 43.25
     + L   89.58
     EC =   16 + 32.83

10.5   Curve #1   Curve #2
        Δ = 12°30'   Δ = 10°56'
        R = 600 ft.   R = 600 ft.
        T_1 = 65.71 T_2 = 57.42
        L_1 = 130.90 L_2 = 114.49

        PI_1 @ 3 + 81.27 PI_2 @   5 + 42.30
        - T_1 65.71 PI_1 @   3 + 81.27
        BC_1 = 3 + 15.56   Diff. =   161.03
        + L_1 1   30.90 $T_1 + T_2$ = -123.13   65.71
        EC_1 = 4 + 46.46 EC_1 to BC_2 = 37.90   + 57.42
                                                                          =123.13

        EC_1 to BC_2 =   37.90
        BC_2 4 + 84.36
        + L_2 1   14.49
        EC_2 5 + 98.85

10.6   T = 300 Tan 6°23'30" = 33.606 m
$$L = 2\pi \times 300 \times \frac{12.783}{360} = 66.933 \text{ m}$$

PI @ 5 + 862.789	Deflection angle for 1m = $\frac{6.3916°}{66.933}$
- T 33.606	
BC = 5 + 829.183	= 0.0954935°
+ L 66.933	Deflections:
EC 5 + 896.116	(1) 10.817 x .0959935 = 1°01'59"
	(2) 20 x .0954935 = 1°54'36"
	(3) 16.116 x .0954935 = 1°32'20"

    BC   5 + 829.183   0°00'00"
               5 + 840   1°01'59"
               5 + 860   2°56'35"
               5 + 880   4°51'11"
    EC   5 + 896.116   6°23'31" ≈ Δ/2

10.7 $E = 500 \frac{1}{\cos 12°12'10"} - 1 = 11.558 \text{ m}$
      M = 500 (1 - Cos 12°12'10") = 11.297 m
      T = 500 Tan 12°12'10" = 108.129 m
      $L = 500 \pi \times \frac{24.40556}{180} = 212.979 \text{ m}$

      PI @   8 + 272.311
      - T   108.129
      BC =   8 + 164.182
      + L   212.979
      EC =   8 + 377.161

10.8   T = 1150 Tan 18°05'15" = 375.60
$$L = 2 \times \pi \times 1150 \times \frac{36.175}{360} = 726.08$$

    PI @   10 + 71.78
    - T   3   75.60
    BC =   6 + 96.18
    + L   7 + 26.08
    EC =   14 + 22.26

	Deflection/ft. = $\frac{\Delta/2}{L}$
	= .0249111°

      BC   6 + 96.18   0°00'00"   Deflections:
              7 + 00   0°05'43"   (1) 3.82 x .0249111 = 0°05'43"
              8 + 00   2°35'11"   (2) 100   x .0249111 = 2°29'28"
              9 + 00   5°04'39"   (3) 22.26 x .0249111 = 0°33'16"
            10 + 00   7°34'07"

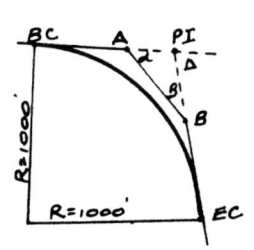

	11 + 00	10°03'35"
	12 + 00	12°33'03"
	13 + 00	15°02'31"
	14 + 00	17°31'59"
EC	14 + 22.26	18°05'15" = Δ/2

10.9 For 10.817m arc, deflection = 1°01'58.6"
    (1) C = 2RSin deflection = 10.816 m
    For 20m arc, deflection = 1°54'36"
    (2) C = 2RSin deflection = 19.996 m
    For 16.116m arc, deflection = 1°32'20"
    (3) C = 2RSin deflection = 16.114 m

10.10 For 3.82 ft. arc, deflection = 0°05'43"
   (1) C = 2RSin deflection = 3.82 ft.
   For 100 ft. arc, deflection = 2°29'28"
   (2) C = 2RSin deflection = 99.97 ft.
   For 22.26 ft. arc, deflection = 0°33'16"
   (3) C = 2RSin deflection = 22.26 ft.

10.11 Δ = 36°10'30" RT, therefore radius of right offset
        curve = 1150 - 50 = 1100 ft. and radius of left
        offset curve = 1150 + 50 = 1200 ft.

Interval	LS.R=1200 ft.	℄.R=1150	RS.R=1100 ft.
BC to 7+00	c=2400Sin0°05'43" = 3.99	c=2300Sin0°05'43" = 3.82	c=2200Sin0°05'43" = 3.65
100 ft.	c=2400Sin2°29'28" = 104.31	c=2300Sin2°29'28" = 99.97	c=2200Sin2°29'28" = 95.62
14+00 to EC	c=2400Sin0°33'16" = 23.22	c=2300Sin0°33'16" = 22.26	c=2200Sin0°33'16" = 21.29

Computation check: At each interval, differences between LS & ℄, and ℄ and RS are equal.

10.12 Measured values:
    AB = 615.27 ft.
    α = 51°31'20"
    β = 32°02'45"

    Δ = α + β = 83°34'05"
    T = R Tan $\frac{\Delta}{2}$

= 1000 Tan 41°47'02"
= 893.60 ft.
Side A,PI = Sin 32°02'45" x $\dfrac{615.27}{\text{Sin } 96°25'55"}$ = 328.53 ft.
Side PI,B = Sin 51°31'20" x $\dfrac{615.27}{\text{Sin } 96°25'55"}$ = 484.71 ft.
To locate BC, Layout (893.60 - 328.53) <u>565.07 back</u> from point A.
To locate EC, Layout (893.60 - 484.71) <u>408.89 forward</u> from point B.

10.13 Sin (0, C.B., V) = OV x $\dfrac{\text{Sin (O, V, C.B.)}}{R}$

R/OV = Cos Δ/2, OV = R/Cos Δ/2
Sin (O,C.B.,V) = R/Cos Δ/2 x Sin (O, V,C.B.)/R
= $\dfrac{\text{Sin (O,V,C.B.)}}{\text{Cos } \Delta/2}$
= $\dfrac{\text{Sin } 32°31'}{\text{Cos } 35°48'}$ = 0.662765

Angle (O,C.B.,V) = 138°29'20"
Angle (C.B.,O,V) = 180 - (138°29'20"+32°31') = 8°59'40"

$\dfrac{R}{\text{Sin } 32°31'}$ = $\dfrac{8.713}{\text{Sin } 8°59'40"}$, R = 29.958 m

10.14 Given R_1 = 200 ft., R_2 = 300 ft.
Δ_1 = 44°26', Δ_2 = 45°18'
Solve for T_1 and T_2 so that the curve
can be located in the field.
t_1 = 200 Tan 22°13' = 81.69 ft.
t_2 = 300 Tan 22°39' = 125.19 ft.
Base of vertex triangle V_1V_2 =
125.19 + 81.69 = 206.88
Δ = 44°26' + 45°18' = 89°44'
Interior Angle V = 90°16'
$\dfrac{V_2V}{\text{Sin } 44°26'}$ = $\dfrac{206.88}{\text{Sin } 90°16'}$, V_2V = 144.83'
$\dfrac{V_1V}{\text{Sin } 45°18'}$ = $\dfrac{206.88}{\text{Sin } 90°16'}$, V_1V = 147.05'
T_1 = t_1 + V_1V = 81.69 + 147.05 = 228.74 ft.
T_2 = t_2 + V_2V = 125.19 + 144.83 = 270.02 ft.

10.15 1.   A = 1.8 - (-3.2) = 5.0
     2.   BVC @ 7 + 25.712 - 50 = 6 + 75.712
         EVC @ 7 + 25.712 + 50 = 7 + 75.712
     3.   Elevation of BVC = 210.440 + (50 x .032) = 212.040
         Elevation of EVC = 210.440 + (50 x .018) = 211.340
     4.   Distance from BVC to low point:
$$X = \frac{-g_1 L}{A} = \frac{3.2 \times 100}{5} = 64.000 m$$
         Low pt @ (6 + 75.712) + 64.000 = 7 + 39.712

STATION		TAN. ELEV.	TANGENT OFFSET $[x/(L/2)]^2 d$	CURVE ELEV.
BVC	6+75.712	212.040	$(0/50)^2$ x .625 = 0.000	212.040
	6+80	211.902	$(4.288/50)^2$ x .625 = .005	211.907
	7+00	211.262	$(24.288/50)^2$ x .625 = .147	211.409
	7+20	210.622	$(44.288/50)^2$ x .625 = .490	211.112
PVI	7+25.712	210.440	$(50/50)^2$ x .625 = 0.625	211.065
Low PT	7+39.712	210.692	$(36/50)^2$ x .625 = .324	211.016
	7+60	211.057	$(15.712/50)^2$ x .625 = .062	211.119
EC	7+75.712	211.340	$(0/50)^2$ x .625 = 0.000	211.340

10.16 1.   A = -1-2.5 = - 3.5
     2.   BVC @ 19 + 00 - 250 = 16 + 50
         EVC @ 19 + 00 + 250 = 21 + 50
     3.   Elevation of BVC = 723.86 - (250 x .025) = 717.61 ft.
         Elevation of EVC = 723.86 - (250 x .01) = 721.36 ft.
     4.   Distance from BVC to summit (high point)
$$X = \frac{-g_1 L}{A} = \frac{-2.5 \times 500}{-3.5} = 357.14 \text{ ft.}$$
         Summit @ (16 + 50) + 357.14 = 20 + 07.14

STATION	TAN. ELEV.	TANGENT OFFSET $[x/(L/2)]^2 d$	CURVE ELEV.
BC16+50	717.61	$(0/250)^2$ x 2.19 = 0	717.61
17+00	718.86	$(50/250)^2$ x 2.19 = .09	718.77
18+00	721.36	$(150/250)^2$ x 2.19 = .79	720.57
PVI19+00	723.86	$(250/250)^2$ x 2.19 = 2.19	721.67
20+00	722.86	$(150/250)^2$ x 2.19 = .79	722.07
Summit 20+07.14	722.78	$(142.86/250)^2$ x 2.19 = .72	722.07

| 21+00 | 721.86 | $(50/250)^2$ x 2.19 = .09 | 721.77 |
| EC21+50 | 721.36 | $(0/250)^2$ x 2.19 = 0 | 721.36 |

10.17
1. A = -1-3 = -4
2. L = KA = 90 x 4 = 360 m
3. BVC @ 0 + 360 - 180 = 0 + 180
   EVC @ 0 + 360 + 180 = 0 + 540
4. Elevation BVC = 156.663 - (180 x .03) = 151.263
   Elevation EVC = 156.663 - (180 x .01) = 154.863
5. High pt. @ - g_1K from BVC
   i.e.: 3 x 90 = 270
   High point station = (0 + 180) + 270 = 0 + 450
   or Highpoint @ g_2K from EVC
   i.e.: 1 x 90 = 90
   High point station = (0 + 540) - 90 = 0 + 450
6. Tangent offset (d) = $\dfrac{KA^2}{800} = \dfrac{90 \times 16}{800}$ = 1.800

STATION	TAN. ELEV.	TANGENT OFFSET $[x/(L/2)]^2 d$	CURVE ELEV.
BVC 0+180	151.263	$(0/180)^2$ x 1.8 = 0	151.263
0+200	151.863	$(20/180)^2$ x 1.8 = .022	151.841
0+250	153.363	$(70/180)^2$ x 1.8 = .272	153.091
0+300	154.863	$(120/180)^2$ x 1.8 = .800	154.063
0+350	156.363	$(170/180)^2$ x 1.8 = 1.606	154.757
PVI 0+360	156.663	1.800	154.863
0+400	156.263	$(140/180)^2$ x 1.8 = 1.089	155.174
High Pt. 0+450	155.763	$(90/180)^2$ x 1.8 = .450	155.313
0+500	155.263	$(40/180)^2$ x 1.8 = .089	155.174
EVC 0+540	154.863	$(0/180)^2$ x 1,8 = 0	154.863

10.18 Ls = 150 ft. (table 10.3)
   Δs = 6°00' (table 10.4)
   R = 716.1972'; p = 1.3085'
   R+p = 717.5057 ft.
   q = 74.9726 ft.
   LT = 100.0575'; ST = 50.0523'; X = 149.84'; Y = 5.23'
   Ts = (R + p) Tan Δ/2 + q = 717.5057 Tan 8°22'+74.9726 = 180.50 ft.
   Δc = Δ - 2Δs = 16°44' - 12°00' = 4°44'; Δc/2 = 2°22'
   Lc = $\dfrac{100\ \Delta c}{D} = \dfrac{100 \times 4.73333°}{8}$ = 59.17 ft.
   θs = Δs/3 = 6°/3 = 2°00'
   Tc = R tan Δc/2 = 716.1972 tan 2°22' = 29.60 ft.

   PI @   11 + 66.18
   - Ts   1 80.50
   T.S. = 9 + 85.68
   +Ls   1 50
   S.C. = 11 + 35.68
   + Lc   59.17
   C.S. = 11 + 94.85
   + Ls   1 50
   S.T. = 13 + 44.85

10.19 Y = Ls Sin θs = 150 x Sin 2° = 5.23 ft.
   X = $\sqrt{Ls^2 - Y^2} = \sqrt{150^2 - 5.23^2}$ = 149.91 ft.
   q = 1/2X = 149.9 1/2 = 74.95 ft.
   p = Y/4 = 1.3075
   LT= Sin 2Δs/3 x Ls/SinΔs = $\dfrac{\text{Sin } 4° \times 150}{\text{Sin } 6°}$ = 100.10 ft.
   St = SinΔs/3 x $\dfrac{Ls}{\text{Sin}\Delta s}$ = $\dfrac{\text{Sin } 2° \times 150}{\text{Sin } 6°}$ = 50.08 ft.

PARAMETER	APPROX. METHOD	TABULAR METHOD	(PROBLEM 10.18)
Y	5.23 ft.	5.23 ft.
X	149.91 ft.	149.84 ft.
q	74.95 ft.	74.97 ft.
p	1.31 ft.	1.31 ft.
LT	100.10 ft.	100.06 ft.
ST	50.08 ft.	50.05 ft.

10.20

STATION		DEFLECTION (minutes)	DEFLECTION (degrees & minutes)
T.S.9+85.68		0	0°00.00'
10+00		1.09'	0°01.09'
10+50	SPIRAL	22.06'	0°22.06'
11+00		69.70'	1°09.70'
S.C.11+35.68		120.00	2°00.00'
S.C.11+35.68		0°	0°00.00'
11+50	CIRCULAR	34.37'	0°34.37'
C.S.11+94.85		142.01'	2°22.01'
C.S.11+94.85		120.00	2°00.00'
12+00		111.90	1°51.90'
12+50	SPIRAL	47.98'	0°47.98'
13+00		10.73'	0°10.73'
S.T.13+44.85		0	0°00.00'

ℓ^2 (θs×60') / (Ls^2) = .005333 ℓ^2

Def^n/ft = 142/59.17 = 2.4000'
14.32 × 2.4000 = 34.37'
34.85 × 2.4000 = 107.64

10.21 For V = 80 Km/h, R = 220 m and two lane road,
A = 125m and e = 0.060 (table 10.5)
For A = 125 and R = 220 m (table 10.7)

Ls = 71.023 m   LT = 47.413 m
X = 70.838 m   ST = 23.733 m
Y = 3.814 m   LC = 70.941 m
q = 35.481 m   Δs = 9°14'54.3"
p = 0.954 m   θs = 3°04'55.6" (precise)

Ts = (R+P) tanΔ/2 + q = 220.954 tan 14°04' + 35.481 = 90.844 m
Δc= Δ - 2Δs = 28°08' - 2 × 9°14'54.3" = 9°38'11"
$\frac{\Delta c}{2}$ = 4°49'06", Lc = $\frac{2\pi R \Delta c}{360}$ = 37.001 m

Tc = R tan Δc/2 = 18.545 m
θs (approx.) = Δs/3 = 3°04'58.1" (error = 2.5 seconds)

```
PI @   1 + 286.441            S.C.= 1 + 266.620
-Ts         90.844           +Lc         37.001
T.S. = 1 + 195.597            C.S. = 1 + 303.621
+Ls         71.023           +Ls         71.023
S.C.   1 + 266.620            S.T. = 1 + 374.644
```

37

10.22

$Y = L_s \sin \theta_s = 71.023 \sin 3°04'55.6" = 3.819$
$X = \sqrt{L_s^2 - Y^2} = \sqrt{71.023^2 - 3.819^2} = 70.920$
$q = X/2 = 35.460$
$p = Y/4 = 0.955$
$LT = \sin 2\Delta s/3 \times L_s/\sin\Delta s = 47.463$ m
$ST = \sin\Delta s/3 \times L_s/\sin\Delta s = 23.766$ m

PARAMETER	APPROXIMATE METHOD	TABULAR METHOD
Y	3.819	3.814
X	70.920	70.838
q	35.460	35.481
p	0.955	0.954
L.T.	47.463	47.413
S.T.	23.766	23.733

10.23

STATION	DEFLECTION (minutes)	DEFLECTION (deg. & min.)	
T.S.1+195.597	0	0°00.00'	$\ell^2 (\theta \times 60)$
1+200	0.71	0°00.71'	(L_s^2)
1+22	21.83	0°21.83'	
1+240	72.28	1°12.28'	$= .036661 \ell^2$
1+260	152.06	2°32.06'	
S.C.1+266.620	184.93	3°04.93'= θs	
S.C.1+266.620	0	0°00.00'	Def²/m = 289.1
1+280	104.54	1°44.54'	37.001
			= 7.8133
			13.380 x 7.8133 = 104.54'
1+300	156.27	4°20.81'	20.000x 7.8133 =156.27'
C.S.1+303.62	128.29	4°49.10 = Δc/2	3.621x 7.8133 = 28.29' / 289.1'
C.S.1+303.621	184.93	3°04.93'= θs	
1+320	109.47	1°49.47'	
1+340	44.00	0°44.00'	
1+360	7.86	0°07.86'	
S.T.1+374.644	0	0°00.00'	

10.24 A = 3-(-1) = 4
   For V = 80 Km/h, K = 30
   For A = 4 and K = 30, L = KA = 120m
   $d = \frac{AL}{800} = \frac{4 \times 120}{800} = 0.600$
   Low point @ g_1K from BVC = 1 x 30 = 30m

STATION	TAN. ELEV.	TANGENT OFFSET=[x/(L/2)]²d	CURVE ELEV.
BVC 1+240	211.000	0	211.000
1+260	210.800	(20/60)² x .600 = .067	210.867
Low Pt 1+270	210.700	(30/60)² x .600 = .150	210.850
1+280	210.600	(40/60)² x .600 = .267	210.867
PVI 1+300	210.400	.600	211.000
1+320	211.000	(40/60)² x .600 = .267	211.267
1+340	211.600	(20/60)² x .600 = .067	211.667
EVC 1+360	212.200	0	212.200

10.25 Lane cross fall @ normal crown = 4 x .02 = .080
   Lane cross fall @ e = .06 = 4 x .06 = .240
   Tangent runout @ 400:1: $\frac{X_1}{.080} = \frac{400}{1}$, X_1 = 32.000 m

T.S. @ 1 + 195.597   S.T. @ 1 + 374.644
-X_1 32   +X_1 32
A   1 + 163.597   A' =   1 + 406.644

Location distance of point C (C') from T.S. (S.T.)
   $\frac{X_2}{.080} = \frac{71.023}{.240}$, X_2 = 23.674

T.S. @ 1 + 195.597   S.T. @ 1 + 374.644
+X_2 23.674   -X_2 23.674
C =   1 + 219.271   C' =   1 + 350.970

STATION	GRADE ℄	LEFT EDGE PAVE. ABOVE/BELOW℄	ELEV.	RIGHT EDGE PAVE. BELOW	ELEV.
1+160	211.800	-.080	211.720	-.080	211.720
A 1+163.597	211.764	-.080	211.684	-.080	211.684
1+180	211.600	-.039	211.561	-.080	211.520
T.S.1+195.597	211.444	.000	211.444	-.080	211.364
1+200	211.400	+.015	211.415	-.080	211.320
C 1+219.271	211.207	+.080	211.287	-.080	211.127
1+220	211.200	+.082	211.282	-.082	211.118
BVC 1+240	211.000	+.150	211.150	-.150	210.850

39

1+260	210.867	+.218	211.084	-.218	210.649
S.C.1+266.620	210.852	+.240	211.092	-.240	210.612
Low Pt.1+270	210.850	+.240	211.090	-.240	210.610
1+280	210.867	+.240	211.109	-.240	210.627
PVI 1+300	211.000	+.240	211.240	-.240	210.760
C.S.1+303.621	211.038	+.240	211.278	-.240	210.798
1+320	211.267	+.185	211.452	-.185	211.082
1+340	211.667	+.117	211.784	-.117	211.550
C' 1+350.970	211.943	+.080	212.023	-.080	211.863
EVC 1+360	212.200	+.049	212.249	-.080	212.120
S.T.1+374.644	212.639	.000	212.639	-.080	212.559
1+380	212.800	-.013	212.787	-.080	212.720
1+400	213.400	-.063	213.337	-.080	213.320
A' 1+406.644	213.599	-.080	213.519	-.080	213.519
1+420	214.000	-.080	213.920	-.080	213.920

CHAPTER 11

11.1		Average Factors	
		Elevation	Scale
A to B; ΔN = -77.773: ΔE = -280.126 Distance AB = 290.722; Brg=S74°29'00.4"W		99997166	99990182
B to C; ΔN = -35.982: ΔE = -223.047 Distance BC = 225.931; BRG=S80°50'09.5"W		99997094	99990175
C to D; ΔN = +2.638: ΔE = +206.021 Distance CD = 206.038m; BRG=N89°15'59.0"E		99997046	99990156
D to A; ΔN = +111.117: ΔE = +297.152 Distance DA = 317.248; Brg=N69°29'50.0"E		99997118	99990163

11.2 A -   4°59'10.4"
     B - 186°21'09.1"
     C -   8°25'49.5"
     D - 160°13'51.0"
         358°118'120.0" = 360°00'00.0"

11.3 Use equations 11-8 and 11-9, or table 11.3 - converted to metres.

MONUMENT	GRID FACTOR	GRID DIST.	GROUND DIST.	COURSE
A	.999874			
B	.999873	290.723	290.759	AB
C	.999872	225.931	225.960	BC
D	.999872	206.038	206.064	CD
A	.999874	317.248	317.288	DA

11.4 Use equation 11-4b (θ = 52.13 D tan ϕ), θ = 32.392D(Km)tanϕ

COURSE	DISTANCE FROM C.M. (Km) (mid point of course)	CONVERGENCE θ	
		θ"	θ
AB	12.164	377.8"	0°06'17.8"
BC	11.912	370.0"	0°06'10.0"
CD	11.904	369.7"	0°06'09.7"
DA	12.155	377.6"	0°06'17.6"

COURSE	GRID BEARING	GEODETIC BEARING
AB	S.74°29'00.4"W.	S.74°35'18.2"W.
BC	S.80°50'09.5"W.	S.80°56'19.5"W.
CD	N.89°15'59.0"E.	N.89°22'08.7"E.
DA	N.69°29'50.0"E.	N.69°36'07.6"E.

11.5 A   4° 59' 10.6"
     B   186° 21' 01.3"
     C   8° 25' 49.2"
     D   160° 13' 58.9"
         358°118'120.0" = 360°00'00.0"

11.6
  7.20.13 pm Aug.10,1995
  +12.00
  ─────────
  19.20.13 LMT
  + 5   Correction for longitude 75° (ie., EST)
  ─────────
  24.20.13

GCT = $0^H 20^M 13^S$ Aug. 11,1995 (UT approx.)

GHA (from Table 11.6)
  Aug.11 @ 24^H 282°58'01.9"
           0^H 281°59'16.1"
                    ────────────
  difference =   0°58'45.8"
                    +360°
                    ────────────
change in 24 hrs. = 360°58'45.8"
Change in **GHA** for $0^H 20^M 13^S$, = $[0^H 20^M 13^S]/24^H$ × 360°58'45.8"
                     = .336944/24 × 360.9794
                     = 5.06791° = 5°04'04.4"
                                 + 281°59'16.1"
                                 ────────────────────
  GHA @ $0^H 20^M 13^S$, Aug.11,1995 = 287°03'20.5"

LHA = GHA - West Longitude (see Figure 11.35)
LHA = 287°03'20.5" - 79°21'00" = 207°42'20.5"

Declination (d) from Table 11.6
Aug.11/95 @ 24^H 89°14'20.46"
          @ 0^H 89°14'20.33"
                    ────────────
Difference in 24 hrs =   00.13"
Difference in $0^H 20^M 13^S$ is insignificant.
Declination (d) Aug.11/91 @ $0^H 20^M 13^S$ = 89°14'20.33"

From equation 11.12, Azimuth :

$$\text{Tan}^{-1} \quad \frac{\text{Sin } 207°42'20.5"}{\text{Sin } 43°47'30" \text{ Cos } 207°42'20.5" - \text{Cos } 43°47'30" \text{ Tan } 89°14'20.33"}$$

Tan Azimuth = +0.00845987, Azimuth = 0°29'05"

Polaris Azimuth = 0°29'05"
+ Field angle   = 43°43'38"
                   ────────────
Azimuth of line 16 - 17 = 44°12'43"

11.7   9.27.30 pm Aug.4,1995
       +12
       21.27.30 LMT
       + 8   Correction for Longitude 120° (PST)

       29.27.30
GCT = $5^H27^M30^S$ Aug.5,1995 (UT approx.)

GHA (from Table 11.6)
            Aug.5 @ 24^H, 277°05'44.2"
                 @ 0^H, 276°07'05.1"
                           ----
       difference   0°58'39.1"
                           +360°
GHA change in 24 hrs = 360°58'39.1"
Change in GHA for $5^H27^M30^S$ = $[5^H27^M30^S]/24^H$ x 360°58'39.1"
                            = 0.2274305 x 360.977528
                            = 82.0972996° = 82°05'50.35"
GHA @ $5^H27^M30^S$ = 276°07'05.1"
                    + 82°05'50.35"

                    = 358°12'55.45"

LHA = GHA - West Longitude (see Figure 11.35)
LHA = 358°12'55.45" - 119°40'00" = 238°32'55.5"

Declination (d) from Table 11.6,
Aug. 5 @ 24^H 89°14'19.58"
       @ 0^H 89°14'19.48"
Difference in 24 hrs=00.10"
Change in d for $5^H27^M30^S$ is $[5^H27^M30^S]/24^H$ x 00.10 = .02
Declination = 89°14'19.48
           +   00.02
           = 89°14'19.50"

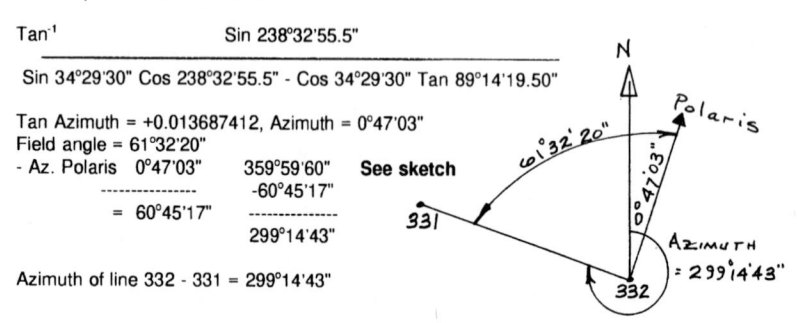

From Equation 11.12 Azimuth:

Tan⁻¹   Sin 238°32'55.5"
       ───
       Sin 34°29'30" Cos 238°32'55.5" - Cos 34°29'30" Tan 89°14'19.50"

Tan Azimuth = +0.013687412, Azimuth = 0°47'03"
Field angle = 61°32'20"
- Az. Polaris 0°47'03"   359°59'60"   **See sketch**
              ---------   -60°45'17"
            = 60°45'17"   ---------
                              299°14'43"

Azimuth of line 332 - 331 = 299°14'43"

43

CHAPTER 13

13.1

STATION	℄ GRADE	STAKE ELEV.	CUT	FILL	CUT	FILL
0+00	472.70	472.60		0.10		0'1 1/4"
1+00	474.02	472.36		1.66		1'8"
2+00	475.34	473.92		1.42		1'5"
3+00	476.66	475.58		1.08		1'1"
4+00	477.98	478.33	0.35		0'4 1/4"	
5+00	479.30	479.77	0.47		0'5 1/2"	
6+00	480.62	480.82	0.20		0'2 3/8"	

13.2

STATION	℄ GRADE	STAKE ELEV.	CUT	FILL
0+00	210.500	210.831	0.331	
0+20	210.365	210.600	0.235	
0+40	210.230	211.307	1.077	
0+60	210.094	210.114	0.020	
0+80	209.959	209.772		0.187
1+00	209.823	209.621		0.202
1+20	209.688	209.308		0.380
1+32.562	209.603	209.400		0.203

13.3   24.5+(3.5 x 3) = 35 ft. from ℄

13.4a 7.45+(2 x 3) = 13.45m from ℄
  b Ditch Invert is 2.67 below original ground; slope stake is 13.45 + (3 x 2.67) = 21.46m from ℄

13.5

STATION	INVERT ELEV.	STAKE ELEV.	CUT	STAKE TO BATTERBOARD FEET	FT/IN
MH8 0+00	360.44	368.75	8.31	5.69	5'8 1/4"
0+50	361.04	368.81	7.77	6.23	6'2 3/4"
1+00	361.64	369.00	7.36	6.64	6'7 3/4"
1+50	362.24	369.77	7.53	6.47	6'5 1/2"
2+00	362.84	370.22	7.38	6.62	6'7 1/2"
MH9 2+40	363.32	371.91	8.59	5.41	5'4 7/8"
	Grade rod = 14'				

13.6

STATION	INVERT ELEV.	STAKE ELEV.	CUT	STAKE TO BATTERBOARD
MH4 0+00	150.666	152.933	2.267	1.733
0+20	150.802	152.991	2.189	1.811
0+40	150.938	153.626	2.688	1.312
0+60	151.074	153.725	2.651	1.349
0+80	151.210	153.888	2.678	1.322
1+00	151.346	153.710	2.364	1.636
MH5 1+15	151.448	153.600	2.152	1.848
Grade rod = 4m				

CHAPTER 14

14.1 For latitude 36°30':
θ(seconds) = 52.13 d TAN ϕ
         = 52.13 x 6 x 0.73996 = 231" = 0°03'51"
θ(seconds) = 32.39 d Tan ϕ
         = 32.39 x 9.66 x .73996 = 231"= 0°03'51"

For latitude 46°30':
θ(seconds) = 52.13 x 6 x 1.05378 = 330" = 0°05'30"
θ(seconds) = 32.39 x 9.66 x 1.05378 = 330" = 0°05'30"

14.2 d = 24 mi.
Y = R(Tan$\Delta\phi$,or Sin$\Delta\phi$), where R = 20,890,000 ft.
$\Delta\phi$ = 40°41' - 40°20' = 0°21'
Y = 20,890,000 x Tan 0°21' = 127,611 ft. = 24.17 mi.
p(feet) = 1.33 dy Tanϕ (Equation 14.8)
        = 1.33 x 24 x 24.17 x Tan (40°41'+ 40°20')/2 = 659 ft.
p(chains) = 0.0202 dy Tan ϕ = 10.0 chains (Equation 14.9)
p(metres) = 0.1565 dy Tanϕ = 201 m (Equation 14.11)
p(metres) = 0.4055 dy Tanϕ = 201m (Equation 14.10)

14.3 Offset for 5 mi. = 3.28 x (5/3)^2 = 9.11 ft. (2.78m)
Offset for 7 mi. = 3.28 x (7/3)^2 = 17.86 ft. (5.44m)
Offset for 10 mi. = 3.28 x (10/5)^2 = 13.12 ft. (4.0m)

14.4 Mean latitude = 46°30'
d = 6 mi., Y = 6 mi.
p(feet) = 1.33 x 6 x 6 Tan 46°30' = 50.4 ft. = 0.0096 mi.
North boundary of Section 6 = 1.0000 - 0.0096 = 0.9904 mi.
Area of Section 6 = 1 mi. x (1.000 + 0.9904)/2 = 0.9952 sq.mi. = 636.9 ac.

14.5 a)   SC   b)   S6
             T21N   1908

    | R6E | R7E | | T20N | R7E |
    | S36 | S31 | | S32 | S33 |
    |-----------| |-------------|
        1908 | T19N | R7E |

         R
    ┌─────┐
 17 │   │ 18
    └─────┘
         X

45

14.6

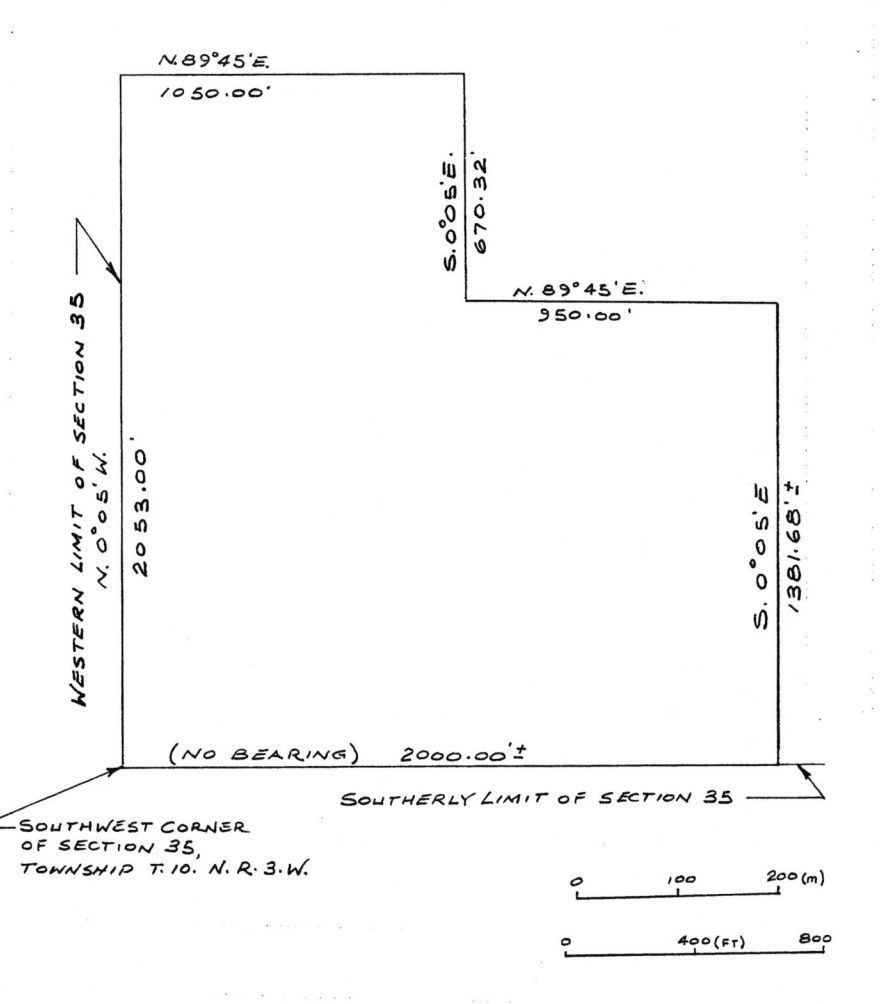

N.89°45'E.
1050.00'

S.0°05'E.
670.32'

N.89°45'E.
950.00'

WESTERN LIMIT OF SECTION 35
N.0°05'W.
2053.00'

S.0°05'E
1381.68'±

(NO BEARING) 2000.00'±

SOUTHERLY LIMIT OF SECTION 35

SOUTHWEST CORNER
OF SECTION 35,
TOWNSHIP T.10. N. R. 3. W.

0   100   200 (m)

0   400 (FT)   800

14.7

New Shoreline

Original Shoreline

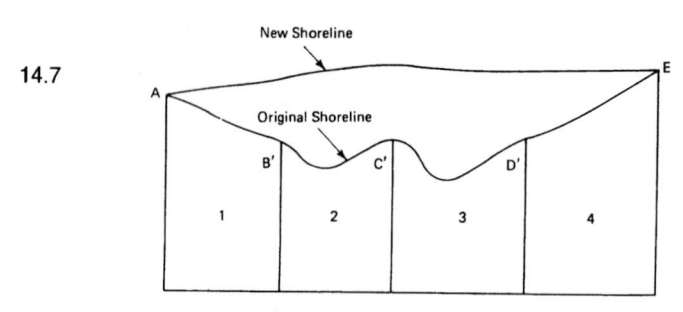

A, B', C', D', E — Lots 1, 2, 3, 4

AE (new shoreline) = 3.85 in.
AE (original shoreline) = 4.70 in.
AB = BC = CD = DE = 3.85/4 = 0.96 in.

Points B, C, and D are established by equal frontages for each of Lots 1, 2, 3, and 4 along the new shoreline.

14.8
That parcel of land consisting of the easterly half of Lot 4 and the westerly half of Lot 5, Registered Plan No. 696, Township of _____, County of _____, Province or State of _____.
(The township number or name, the county name, and the province or state are always stated).

14.9
The difference in bearing to N1°50'W from N2°00'W will reduce the length of the line VY.
Reduction in length = 120.00 x sin (2°00' - 1°50')
      = 120.00 x sin 0°10' (.00291)
      = 0.35 ft.
∴ New distance VY = 240.00 = 0.35 = 239.65 ft.
∴ Rear dimensions of each lot = 239.65/6 = 39.94 ft.

14.10
(a) Length of rear boundary is reduced by 5.04 - 4.97 = 0.07 ft.
 Therefore, new length = 25.06 - 0.07 = 24.99 ft. The bearing is the same.
(b) Change in bearing is calculated as follows:
 tan (bearing change) = $\frac{0.07}{26.75}$ = 0.00262

 ∴ Bearing change = 0°09'00"
 ∴ New bearing = N10°12'00"E + 0°09'00" = N10°21'00"E
 The distance is unchanged.

CHAPTER 15
15.1

	DEPTH USING ECHO SOUNDER	TIDAL* HEIGHT	WEATHER** CONDITIONS	REDUCED SOUNDING
a)	10.2 m	1.5 m	- 0.3 m	9.0 m
b)	15.7 m	- 0.8 m	+ 0.5 m	16.0 m
c)	32.7 ft.	+ 3.8 ft.	- 1.4 ft.	30.3 ft.
d)	50.2 ft.	- 2.3 ft.	- 1.8 ft.	54.3 ft.

* Tidal height above (+) or below (-) chart datum
** Error due to weather conditions above (+) or below (-) tidal water level at time of sounding.

Maximum depth where there is no gap in soundings, if sounding lines are 30m apart.

Using formula in text, Depth $_{Max}$ = 1.85 x 30 m = <u>55.5 m</u>

15.2

SOUNDING LOCATIONS ON OR NEAR DANGER CIRCLE	SHORE STATIONS TO BE USED FOR POSITION-FIXING
1	3, 4, 5
2	3, 4, 6
3	3, 4, 6
4	3, 5, 7
5	3, 5, 7
6	4, 6, 7

For Scale of 1:5,000, Spacing between Line = 50 m and between Fixes on each Line = 125 m

Shoreline Stations Placed to Satisfy Sextant Angle Requirements from Each Fix; Also in Linear Pattern along Shoreline to Minimize Occurence of Danger Circles in Offshore Sounding Area.

Overall Horizontal Control System: May be Established Provincial/State Coordinate System Monuments or Traverse set up for Particular Hydrological Survey being Conducted. Shoreline Stations Tied to Control System and Coordinated to Required Accuracy.

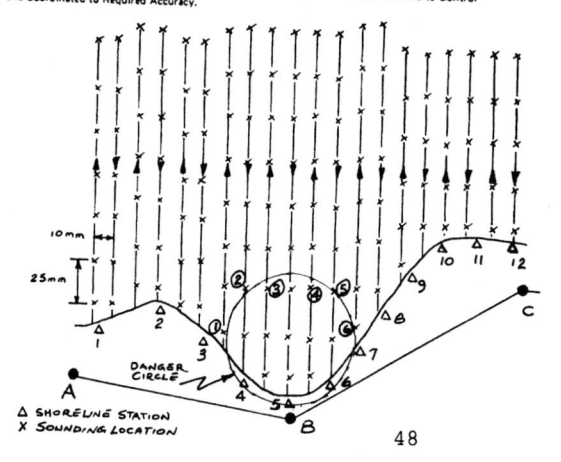

15.3 DEPTH CONTOURS

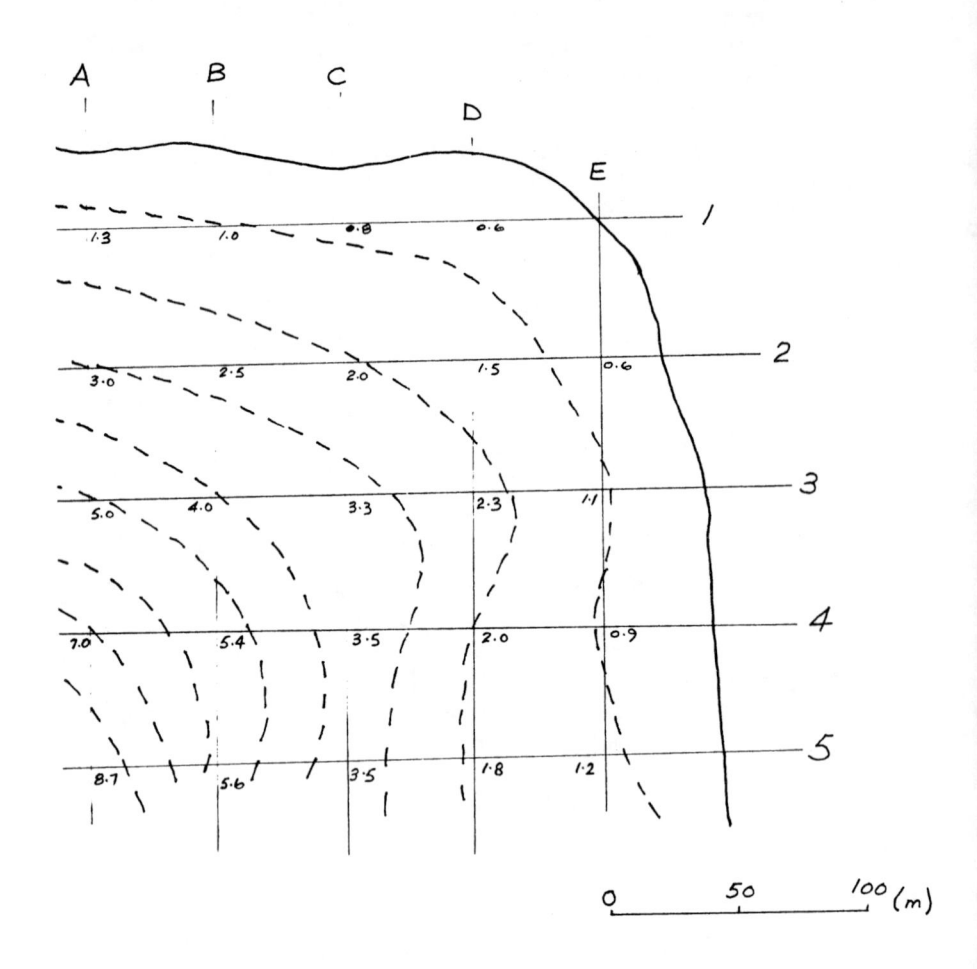

15.4
Maximum depth where there are no gaps in soundings, if sounding lines are 30m apart.

Using formula in text:

Depth $_{Max.}$ = 1.85 x 30m = 55.5m

15.5

SCALE	DISTANCE BETWEEN SOUNDING LINES	DISTANCE BETWEEN FIXES
a) 1:2000	2000 x 0.01m = 20m	1000 x 0.025m = 50m
b) 1 in.=100 ft.	100ft./in. x 0.4in.=40ft.	100ft./in. x 1.0in.=100ft.

15.6
CONDITIONS: Accuracy requirements ± 0.25m
　　　　　　　Max. Distance offshore 10 km

For the conditions listed above

A <u>continuous wave</u> electromagnetic position-fixing system should be used because it can fulfil the required accuracy; rain and fog will not affect the accuracy; and neither will atmospheric conditions such as temperature inversions and water vapour.

CHAPTER 16

16.1 a) H = SR.f = 20,000 × .153 = 3060m
Altitude = 3060 + 180 = 3240m

b) SR = 20,000 × 12 = 240,000
H = SR.f = 240,000 × 6.022/12 = 120,440 ft.
Altitude = 120,400 + 520 = 120,920 ft.

16.2 a) Photo scale = 23.07 × 1:50,000/4.75 = 1:10,295

b) Photo scale = 6.20 × 1:100,000/1.85 = 1:29,839

16.3 a) No. of photos required = 30 × 45/1 $(10,000)^2/(30,000)^2$ = 150

b) No. of photos required = 15 × 33/0.4 $(10,000)^2/(15,000)^2$ = 550

c) SR is 1:500 × 12 or 1:6000
No. of photos required = 10 × 47/0.4 $(10,000)^2/(6,000)^2$ = 3265

16.4 Assume that the negative size is 9 in. (228mm) square, the normal size.

a) Dimension across flight line = 10,000 × .228 = 2280m
or 10,000 × 0.75 = 7,500 ft.
Dimension along flight line = 2280 × 0.60 = 1368m
or 7,500 × 0.60 = 4500 ft.

b) SR is 1:400 × 12 or 1:4,800
Dimension across flight line = 4,800 × .228 = 1094m
or 4,800 × 0.75 = 3,600 ft.
Dimension along flight line = 1094 × 0.60 = 656 m
or 3,600 × 0.60 = 2160 ft.

16.5 a) (i) Ground speed of aircraft = 350 km/hr
= 350 × 1000 / 60 × 60 = 97.2 m/s
∴ camera would move 97.2 × 1/100 = .972m during exposure.

(ii) Ground speed of aircraft = 350 km/hr = 97.2 m/s
∴ camera would move 97.2 × 1/1000 = .097m

(iii) Ground speed of aircraft = 200 miles per hour
= 200 × 5280 / 60 × 60 = 293 ft./sec.
∴ camera would move 293 × 1/500 = 0.59 ft. (018m)

b) The "image motion" during the exposure is least in situation (ii). Therefore, the resolution of an airphoto taken under these conditions would be higher, assuming that all other conditions (weather, film types) are equal.

16.6 A longer focal length decreases the relief displacement. The explanation for this is evident from examining Figure 16.8 b). A longer focal length increases the distance from C to the focal plane. Therefore, the relief displacement, such as bb' and cc' is reduced.

16.7 A wide-angle lens increases relief displacement. As illustrated in Figure 16.8 b), $\angle ACE$ is the lens angle. If this angle is increased, the effect caused by a wide-angle lens, the distance from point "C" to the ground is decreased. Therefore, the radial displacement, such as bb' and cc' is increased.

16.8 The best alternative is d), for the following reasons. The major consideration is relief displacement.
- (i) The focal length is longer than a) and b), thus reducing the radial distortion.
- (ii) The scale of d) is 1 in. = 2000 ft., a scale ratio (SR) of 1:24,000. This is the highest scale ratio of all the alternatives. Reference to Figure 16.8 b) will illustrate that the higher the scale ratio, the greater the distance from C to the ground, and the less the radial distortion.

16.9 On the ground, the dimensions of the area covered by one photograph are 70mm x 500 = 35,000mm = 350m (1150 ft.). For a forward overlap of 60%, the distance along the direction of the flight line = 350 x 0.60 = 210m. Therefore, the "new" area to be photographed between successive exposures is 350 - 210 = 140m. Hence, an exposure must be taken every 140m along the flight line. The ground speed of the aircraft = 160 km/hr = 160 x 1000/(60 x 60) = 44.4 m/s. Time between exposures thus equals 140/44.4 = 3.15, say 3.2 seconds.

16.10 a) The site drainage is very poor. Otherwise, the drainage improvements would not be necessary.
b) The watershed boundaries would have to be determined using stereoscopic viewing of the airphotos, possibly combined with selective field levelling procedures. The location of the original stream banks would have to be accurately determined. The most economical means of achieving this would be the acquisition and use of airphotos taken before the improvements were made.

The channel straightenings, particularly in the upper reaches in Figure 16.21 are located along property lines having different owners on each side. The surveyor would prepare a drainage plan showing the following:
- (i) Location of original natural channels and proposed channel straightenings.
- (ii) Areas of land to be lost through excavation on both sides of the property line.
- (iii) Drainage area sizes on each side of the channel straightening, indicating the relative benefits to each owner.

The drainage plan can then be used for logical negotiations between or among the affected owners regarding all cost items. Both increased drainage benefits can be costed, as well as land lost resulting from excavation or land lost or gained through shifting the channel.

16.11 Yes, because no significant gullies have developed along the sides of the drainline in spite of the steep slopes.

16.12 a) A severe land slide has occurred previously to the right of the area around H.4. The scarp runs generally from G, 7 to I, 5. As the area around H.4 is subject to future soil failures, any survey monumentation placed in this area cannot be considered to be permanent. Therefore, the monument location should be referenced to two or more witness monuments set along the road from J, 1 to G, 4.

b) The degree and type of vegetation regeneration around H, 5 to H, 9 indicates that the landslide occurred between 10 and 15 years before the airphoto was taken.

16.13 a) Wind erosion of find grained, poorly graded sands.
b) 81°30' or 98°30' depending on which side the angle is measured.
c) Drainage: regionally crudely parallel topography: undulating, relief less than 15m. Local erosion: rare deflation basins. Parent materials: coarse-grained, crudely stratified.

16.14 a) All bare soils are highly reflective. Therefore, as the film records only reflected light, the soils appear to have a very light tone.
b) The most significant feature regarding resurveys in this area is the distinct line of large trees running from F, 23 to I, 22. The straight line of trees indicates that it represents an old, well-established property boundary which is likely an original township subdivision line. The surveyor would expect to find evidence of an old boundary fence, probably stapled to the trees within this wooded line. The fence location would determine the precise boundary line.

16.15 Drainage: drainageless (internal)
Topography: crested, uplands asymmetric, lee slope 34°
Local erosion: gullies absent, sand smears.
Parent materials: eolian sand
Soil profile: none (indicated by sparsity of vegetation on highs).

16.16 a) G.5, 6 and I.5, 4 (5.5 indicates halfway between G and H; I.5 indicates halfway between I and J).
b) Rectangular.

16.17 a) K, 9.5
b) Drainage: deranged.
Topography: rolling; random; some basins; relief less than 60m.

Location erosion: (Disagrees with table as there are no gullies. This indicates low clay content in the soil).
Parent materials: glacial drift (heterogeneous)

16.18 a) I, 30 and N, 39
b) Forest fire, known as an "old burn".

16.19 a) Organics (Pt, using Unified soil classification system)
b) The conditions are:
   (i) Deep organic soils, thus difficulties in traversing the terrain, establishing stable instrument set ups, and monumentation.
   (ii) Dense tree growth up to 8-10 ft. high, thus requiring considerable labour in clearing lines.
   (iii) Wide river to be crossed, thus involving tri-angulation or trilateration; also affects equipment requirements and involves boat rental.

16.20 Lineal texture is due to plough lines related to agricultural land uses.

16.21 a) Outwash surfaces. The pattern fits the pattern elements given in Table 16.3.
b) Schist hills. The pattern fits the pattern elements given in table 16.3.

16.22 Difference in tone (light on right, darker on left) reflecting different agricultural crops in the past. This is a subtle, yet important, indication of property line evidence.

SURVEYING AND ADVANCED SURVEYING

REVIEW SUMMARY

FOR USE WITH

SURVEYING PRINCIPLES AND APPLICATIONS, 4TH ED

BARRY KAVANAGH, 1996

NOTE: These review summary sheets are included for those faculty who expressed an interest in obtaining review guides suitable for copying for student distribution.

CONTENTS

PART ONE   **SURVEYING**

INTRODUCTION	57
TAPING	57
LEVELLING	58
ANGLES/TRANSITS/THEODOLITES	59
TRAVERSE COMPUTATIONS	60
TOPOGRAPHIC SURVEYS AND DRAWINGS	61

PART TWO   **ADVANCED SURVEYING**

HIGHWAY CURVES	62
HIGHWAY CONSTRUCTION SURVEYS	63
MUNICIPAL CONSTRUCTION SURVEYS	63
PIPELINE AND SEWER CONSTRUCTION SURVEYS	64
ELECTRONIC SURVEYING MEASUREMENT	65
CULVERT AND BRIDGE SURVEYS	66
BUILDING SURVEYS	66
CONTROL SURVEYS	67
GLOBAL POSITIONING SURVEYS	68
QUANTITY AND FINAL SURVEYS	69
HIGHWAY CURVES (continued)	69
LAND SURVEYING	70
HYDROGRAPHIC SURVEYS	71
PHOTOGRAMMETRY	71

SURVEYING REVIEW SUMMARY

PAGE	CHAPTER 1 INTRODUCTION
	Definitions
1	Surveying
3	Engineering Surveying
1	Plane Surveying
1	Geodetic Surveying
2	Preliminary Surveys
3	Layout Surveys
3	Control Surveys
2	Construction Surveys
4	Hydrographic Surveys
4	Topographic Surveys
	Topics
11	Location Techniques: Rectangular, Polar, Intersection
10	Units of Measurement
11	Precision, Accuracy, Errors, Mistakes
13	Stationing
6	Geographic Reference
14	Field Notes
16	Evolution of Surveying
	CHAPTER 2 TAPING
	Topics
22	Methods of Linear Measurement: Pacing, Odometer, Stadia,
23	EDM, Subtense, Tapes (steel & cloth), Gunter's Chain
29	Taping Accessories: Hand level, Plumb bob, etc.
32	Taping Techniques
28	Standard Conditions for Steel Tape Use
36	Taping Corrections (Systematic Errors)
36	Slope: angle, vertical side, gradient (%)
40	Erroneous Tape length
41	Temperature
42	Tension, Sag (including Normal Tension)
46,48	Random Taping Errors
49	"Ordinary" Taping Precision
50	Mistakes in Taping
51	Problems

SURVEYING REVIEW SUMMARY

PAGE **CHAPTER 3 LEVELLING**

<u>Definitions</u>

54	Levelling
54	Elevation
54	Mean Sea Level (MSL)
54	Vertical Line
54	Horizontal Line
54	Level Line
54	Differential Levelling
70	Benchmark (BM)
72	Temporary Benchmark (TBM)
72	Turning Point (TP)
72	Backsight (BS)
72	Height of Instrument (HI)
72	Foresight (FS)
72	Intermediate Sight (IS)

<u>Topics</u>

56	Curvature & Refraction (c+r)
	Types of Levels
58	Dumpy (including set-up technique)
63	Tilting (including set-up technique)
65	Automatic (including set-up technique)
67	Digital Level and Bar Code Rod
69	Levelling Rods
75	Techniques of levelling
78	Benchmark Levelling (Vertical Control Surveys)
79	Profile and Cross sectioning Levelling
84	Reciprocal Levelling
85	Peg Test
87	Three-Wire Levelling
89	Trigonometric Levelling
90	Level Loop Adjustments
92	Suggestions for the Rod/Instrument Operator
93	Mistakes in Levelling
94	Problems

SURVEYING REVIEW SUMMARY

PAGE	ANGLES, TRANSITS/THEODOLITES CHAPTER 4, 5
	<u>Definitions</u>
127	Transit/Theodolite
105	Interior Angle (& polygon closure)
106	Deflection Angle
106	Exterior Angle
103	Zenith Angle
103	Vertical Angle
103	Nadir Angle
127	Alidade
130	Circle Assembly
134	Levelling Head Assembly
	<u>Topics</u>
103	General
103	Vertical Angles
103	Horizontal Angles
127	Transit components
130	Circle and vernier readings
133	Telescope
134	Transit set-up
135	Measuring angles by repetition
143	Repeating optical theodolites
137	Instrument adjustments, Geometry of the Transit
152	Direction optical theodolites
145	Theodolite set-up (optical plummets)
147	Electronic theodolites
155	Laying off angles
156	Prolonging a straight line (Double-centering)
157	Interlining (Balancing in)
158	Intersection of two straight lines
159	Triangulation
159	Prolonging a line past (through) obstacles

SURVEYING REVIEW SUMMARY

PAGE   **TRAVERSE COMPUTATIONS**
              CHAPTER 4, 6

<u>Definitions</u>
103	Geographic meridian
104	Grid meridian
162	Open Traverse
103	Closed Traverse
108	Bearing
107	Azimuth
119	Magnetic declination
120	Isogonic line
166	Latitude
166	Departure
173	Linear error of closure
173	Accuracy ratio
177	Compass rule
205	Meridian distance

<u>Topics</u>
164	Balancing Angles
116	Bearing computations
111	Azimuth computations
108	Bearing/azimuth relationships
119	Magnetic direction
167	Computation of latitudes and departures
173	Computation of linear error of closure
173	Computation of accuracy ratio
175	Traverse precision and accuracy
176	Traverse adjustments
178	Effects of traverse adjustments on original data
180	Omitted measurements
184	Rectangular coordinates computation
189	Geometry of rectangular coordinates
202	Area computations using coordinates
205	Area computations using DMD's
211	Problems

SURVEYING REVIEW SUMMARY

PAGE   **CHAPTER 7 TOPOGRAPHIC SURVEYS and DRAWING**

<u>Definitions</u>

Page	
2	Preliminary/Pre-engineering Surveys
11	Rectangular tie-in
11	Polar tie-in
218	Split-line field notes
221	hi (height of instrument)
221	Rod interval
293	Contour
298	Ridge line
298	Valley line
302	Cross section
301	Profile

<u>Topics</u>

Page	
215	Precision of topographic measurements
214, 290	Map and plan scales
291	Title blocks
292	Drawing sizes
216	Tie-ins by right-angle offsets & field notes
219	Cross sections and profiles
221	Stadia principles
223	Stadia measurement practice and field notes
235	Self-reducing stadia theodolite
293	Contours—construction and interpretation
314	Problems

ADVANCED SURVEYING REVIEW SUMMARY

PAGE	CHAPTER 10 HIGHWAY CURVES
318	

	Definitions
318	Right of Way
319	Point of Intersection (PI)
319	Beginning of Curve (BC), or Point of Curve (PC)
319	End of Curve (EC), or Point of Tangency (PT)
320	Tangent (T), T=RTanΔ/2
320	Chord (C), C=2RSinΔ/2
320	Mid Ordinate (M), M=R(1-CosΔ/2)
321	External (E), E=R(SecΔ/2-1)
321	Arc Length (L), L=2πRΔ/360
322	Degree of Curve (D), D=5729.58/R
322	Arc Length (L), L=100Δ/D

	Topics
318	Route Surveys
318	Circular Curve Geometry
322	Curve Stationing
325	Circular Curve Deflections
327	Chord Calculations
328	Metric Considerations
329	Field Procedures
330	Moving up on the Curve
331	Offset Curves
339	Compound Curves
340	Reverse Curves
379	Problems

Parabolic (Vertical) Curves, Spiral Curves, and Superelevation will often be covered in Highway Technology, or Highway Design subjects. See page 69.

ADVANCED SURVEYING REVIEW SUMMARY

PAGE	HIGHWAY CONSTRUCTION SURVEYS
	Definitions (also, see Glossary)
482	Grade
472	Arterial
472	Collector
472	Freeway
472	Local
487	Clearing
487	Grubbing
487	Stripping
494	Batter Boards
	Topics
483	Highway Classification and Design
487	Slope stakes
491	Layout for Line and Grade
493	Grade Transfer
495	Ditch Construction

472	MUNICIPAL CONSTRUCTION SURVEYS
	Definitions
479	Cut
479	Fill
472	Crown of Road
472	Boulevard
473	Street line
	Topics
472	Classification of Roads
472	Road Allowances
472	Cross Sections
473	Plan and Profile
474	Establishing Centre-line
478	Establishing Offset Lines
479	Construction Grades For a Curbed Street
483	Street Intersections
485	Sidewalk Construction Survey
526	Problems

ADVANCED SURVEYING REVIEW SUMMARY

PAGE   **PIPELINE AND SEWER CONSTRUCTION SURVEYS**

<u>Definitions</u>

497	Crown of Pipe
496	Sanitary Sewer
496	Storm Sewer
497	Springline
497	Invert
497	Manhole
503	Catchbasin
500	Grade Rod
500	Batter Boards
524	Precise Optical Plummet

<u>Topics</u>

504	Pipeline Construction Surveys
496	Sewer Construction Surveys, General
497	Layout for Line and Grade
502	Laser Alignment
503	Catchbasin Layout
506	Tunnel Construction Surveys
526	Problems

ADVANCED SURVEYING REVIEW SUMMARY

PAGE   **ELECTRONIC SURVEYING MEASUREMENT**

<u>Definitions</u>

241	EDM
242	Total Station
241	Infrared
241	Microwave
246	Forced-centering Equipment
257	Data Collector (Electronic Field Book)
278	Digitizer
284	Layerized Data Base

<u>Topics</u>

241	General
242	Principles of EDM measurement
246	EDM Characteristics
245	Atmospheric Corrections
246	Prisms, prism constant
249	EDM Operation
252	Geometry of EDM Measurements
255	EDM Without Reflecting Prisms
257	Total Stations
265	Field Procedures for Total Stations
273	Construction Layout Using Total Stations
275	Overview of Computerized Surveying Systems
281	Geographic Information Systems
282	The New Information Utility
285	Problems

ADVANCED SURVEYING REVIEW SUMMARY

PAGE   **CULVERT, BRIDGE AND BUILDING SURVEYS**

Definitions

520	Skew Number
522	Open Footing Culvert
522	Box Culvert
522	Types of Culverts

Topics

521	Culvert Construction Surveys
509	Bridge Construction, General
510	Contract drawings
511	Layout Computations
516	Offset distance Computations
517	Dimension Verification
519	Grades-Vertical control
520	Footing Cross Sections
521	Building Construction

ADVANCED SURVEYING REVIEW SUMMARY

PAGE   **CONTROL SURVEYS**

<u>Definitions</u>

397	Elevation Factor
398	Scale Factor
399	Grid Factor
399	Ground Distance
401	Grid Distance
401	Convergence
417	Reference Azimuth Point (RAP)
393	Triangulation
385	Trilateration
414	Strength of Figure
419	"Close the horizon"
434	Time

<u>Topics</u>

382	General
Appendix	Horizontal Control Specifications
Appendix	Vertical Control Specifications
422	Positional Accuracies (ISO)
Appendix	Specifications for Short Lines
386	Universal Transverse Mercator Grid System (UTM)
392	State Plane Coordinate Grid System
397	Ground/Grid Conversion
401	Convergence
415	Project Control
382	Horizontal Control
90	Level Loop Adjustments
426	Control Survey Markers
432	Observation on Polaris
434	Time
437	Procedure for Observing Polaris
441	Computations for Azimuth
443	Direction of a Line by Gyrotheodolite
445	Problems

ADVANCED SURVEYING REVIEW SUMMARY

PAGE   GLOBAL POSITIONING SYSTEMS (GPS)

Definitions

446	NAVSTAR
446	Doppler effect
454	Differential mode
454	GDOP
454	PDOP
465	Geoid height
465	Ellipsoid height
457	Real time positioning
461	Sky plot

Topics

446	General
450	Receivers
453	Satellites
454	Errors
455	Static surveys
456	Kinematic surveys
457	GPS applications
458	Survey planning
459	Initial ambiguity resolution
464	Vertical positioning
465	Geoid modelling

ADVANCED SURVEYING REVIEW SUMMARY

PAGE   **QUANTITY AND FINAL SURVEYS**

Topics

311	Area Computations
309	Trapezoidal Technique
310	Simpson's 1/3 Rule
311	Areas By Graphical Analysis
302	Construction Volumes
304	Cross Sections, End Areas,
307	Prismoidal Formula
471	Final (As-Built) Surveys

HIGHWAY CURVES (continued)

Topics

341	Vertical Curves, General
343	Geometric Properties of the Parabola
344	Computation of the High or Low Point on a Vertical curve
345	Elevation Computation Procedures
349	Design Considerations
350	Spiral Curves, General
358	Spiral Curve Computations
360	Spiral Layout Procedure Summary
367	Approximate Solution for Spirals
369	Superelevation, General
369	Superelevation Design
379	Problems

ADVANCED SURVEYING REVIEW SUMMARY

PAGE   **LAND SURVEYING**

Definitions
528	Metes and Bounds
531	Initial Point
532	Principal meridian
547	Bearing Tree
546	Blazes
551	Adverse Possession
551	Alluvium
551	Avulsion
551	Deed Description
552	Fee Simple
552	High water Mark
552	Mortgage
552	Patent

Topics
527	General
528	Historical summary
530	Public Land surveys
533	Convergence and meridians
536	Baselines—Parallels of Latitude
539	Township Boundaries
545	Corner Monumentation and Line Marking
551	Property Conveyance
552	Deed Descriptions
557	Rural land Surveys
565	Urban land surveys
571	Problems

ADVANCED SURVEYING REVIEW SUMMARY

PAGE	HYDROGRAPHIC SURVEYS
	<u>Definitions</u>
574	Sounding
576	Isonified Area
587	Swells
593	Sweeping
590	Multi-path
591	Triangle of uncertainty
	<u>Topics</u>
573	General
574	Planning
575	Depth and Tidal Measurements
580	Position Fixing
595	Sounding Plan
599	Problems

PHOTOGRAMMETRY

<u>Definitions</u>
601	Photogrammetry
608	Altitude = flying height + mean datum
610	Scale Ratio, SR = H/f
611	Photo scale = (map scale/map distance) x photo distance
611	Fiducial marks
612	Principal Point
624	Parallax
625	Difference in Elevation, dh = dph/dp+b
626	Floating Mark

<u>Topics</u>
601	Introduction
601	Camera Systems
608	Photographic Scale
610	Flying Height and Altitude
611	Relief Displacement
613	Flight Lines and Photograph Overlap
615	Ground Control for Mapping
619	Mosaics
621	Stereoscopic Viewing and Parallax
626	Photogrammetric Plotting Instruments
633	Orthophotos
640	Problems